Thermal Transport Characteristics of Phase Change Materials and Nanofluids

This book provides detailed information related to nanofluids, synthesis and preparation, morphologies of nanoparticles, selection of base fluids and thermophysical properties of nanofluids. The advantage of various conduits, the improvement of the heat transfer performance of phase change materials (PCMs), and the base PCMs for diverse applications are also discussed. Crucial difficulties like stability, aggregation, and clogging of nanoparticles are detailed including factors like the size, shape, and motion of nanoparticles that influence the heat transfer performance of nanofluids. Challenges, applications, and scope of the future works in the subject area are included.

Features:

- Covers heat transfer techniques in utilization of base fluids application of phase change materials (PCMs)
- Describes preparation and characterization of nanofluids and nano-based PCMs
- Explains how nanoscience can be utilized in heat transfer studies
- Reviews conventional heat transfer fluids

This volume is aimed at graduate students and researchers in thermal engineering, heat transfer, material science and engineering, and heat transfer enhancement.

Emerging Materials and Technologies

Series Editor: Boris I. Kharissov

For more information about this series, please visit: www.routledge.com/Emerging-Materials-and-Technologies/book-series/CRCEMT

Thermal Transport Characteristics of Phase Change Materials and Nanofluids

S. Harikrishnan and A.D. Dhass

CRC Press
Taylor & Francis Group
Boca Raton London New York

CRC Press is an imprint of the
Taylor & Francis Group, an **informa** business

First edition published 2023
by CRC Press
6000 Broken Sound Parkway NW, Suite 300, Boca Raton, FL 33487-2742

and by CRC Press
4 Park Square, Milton Park, Abingdon, Oxon, OX14 4RN

CRC Press is an imprint of Taylor & Francis Group, LLC

© 2023 S. Harikrishnan and A.D. Dhass

Library of Congress Cataloging-in-Publication Data
Names: Harikrishnan, S. (Santhanam), author. | Dhass, A. D., author.
Title: Thermal transport characteristics of nanofluids and phase change materials /
    S. Harikrishnan and A.D. Dhass.
Description: First edition. | Boca Raton : CRC Press, 2023. | Series: Emerging materials and
    technologies | Includes bibliographical references and index.
Subjects: LCSH: Nanofluids—Thermal properties. | Change of state (Physics) | Heat storage.
Classification: LCC TJ853.4.M53 H37 2023 (print) | LCC TJ853.4.M53 (ebook) |
    DDC 620.1/06—dc23/eng/20220801
LC record available at https://lccn.loc.gov/2022025620
LC ebook record available at https://lccn.loc.gov/2022025621

ISBN: 978-0-367-75704-5 (hbk)
ISBN: 978-0-367-75716-8 (pbk)
ISBN: 978-1-003-16363-3 (ebk)

DOI: 10.1201/9781003163633

Typeset in Times
by Apex CoVantage, LLC

Contents

Author Biographies

S. Harikrishnan received his B.E. in Electrical & Electronics Engineering from the University of Madras, in 2002. Also, he received his M.E. in Refrigeration & Air Conditioning Engineering and a Ph.D. in Faculty of Mechanical Engineering from Anna University, in the years 2007 and 2015, respectively. At present, he is working as Professor & Head in the Department of Mechanical Engineering at Kings Engineering College, Chennai, India. He is an active researcher in the fields of phase change materials, nanofluids, and supercapacitors. He has published many papers in refereed journals, and in conference proceedings. He is one of the recognized research supervisors at the Centre for Research, Anna University, Chennai. He guided two research scholars successfully in the fields of phase change materials and nanofluids. He is also one of the editorial team members in two Scopus indexed journals and one in the *SCI-E* journal. He served as the panel session chair, reviewer, and Managing Guest Editor for international conference proceedings. He has teaching and research experiences of 18 and 12 years, respectively.

A.D. Dhass received his B.E., M.E., and Ph.D. degrees in Electrical and Electronics Engineering, Energy Engineering, and Solar Energy in 2005, 2009, and 2018 respectively, from Anna University, Chennai, India. At present, he is working as Assistant Professor at the Indus Institute of Technology and Engineering (IITE), Indus University, Ahmedabad, Gujarat, India. His areas of research include solar, heat transfer, and phase change materials.

Preface

The main emphasis of the book is to provide adequate information about the importance of the use of phase change materials and nanofluids. This book starts with the importance of thermal energy storage systems and why latent heat storage (LHS) systems are widely preferred over sensible heat storage systems. Then, it follows the basics, functions, and classification of the phase change materials (PCMs) employed in LHS systems. Also, various techniques are recommended to improve the performances of the PCMs as they possess very low thermal conductivity. Amongst the availability of different enhancement techniques, nanomaterials-enhanced PCMs are explicitly discussed. Further, different synthesis and characterization methods of nanomaterials are stated in detail. The size, structure, and concentration of the nanomaterials play a crucial role in deciding the degree of enhancement. In regard to these facts, the preparation and thermal transport properties of nanomaterials-enhanced PCMs for cooling and heating applications are discussed elaborately.

Apart from PCMs, nanofluids are included in the book and it could be fruitful to gain sufficient information on how nanofluids are different from PCMs. Herein, different properties of the nanofluids are discussed and especially, thermophysical properties. The effect of size, structure, and concentration of the nanomaterials on the thermal properties are reported. Further, single nanofluids and hybrid nanofluids are presented with respect to concentration and types of nanomaterials and finally, this book ends with the applications of PCMs and nanofluids.

I hope this book will be beneficial to graduates, postgraduates, and research scholars. Also, suggestions and comments from the readers are highly appreciable and they could be taken into account during the publication of the next edition.

S. Harikrishnan
A.D. Dhass

Acknowledgments

S. Harikrishnan: At the outset, I would like to express my sincere gratitude to Dr.S.Kalaiselvam since I acquired enough knowledge on the phase change materials and nanofluids when I did my research work under his supervision. Next, I am grateful to my father (Mr.E.Santhanam), mother (Mrs.S.Selvambal), wife (Dr.P.Anandhi), and son (Mr.Mohithram), for their never-ending love and support that made me complete this book successfully.

A.D. Dhass: I am grateful to my dear friends Mr.J.N.Bala Ganesh, Mr.T.J.Dorairaj, and my family members including my father (Mr.A.Desappan), mother (Mrs.A.D.Meenakshi), wife (Mrs.P.K.Aparna), daughter (Miss.A.D.Tanishka), and son (Mr.A.D.Jai Krishnan), for their never-ending love and support that made me complete this book successfully.

S. Harikrishnan
A.D. Dhass

1 Introduction

CONTENTS

1.1 THERMAL ENERGY STORAGE

Energy demand has increased in several countries and as a consequence of recent advances in energy consumptions. It exacerbates the mismatch between demand and supply in energy, particularly during periods of peak load demand. In order to improve electricity generation it is necessary to establish new conventional generating stations or renewable energy sources (solar, wind, biomass, etc.). Thus, the gap between energy demand and supply may be conduits. But in the case of conventional power stations, they require fossil fuels. A further source of electricity generation is renewable energy, although it is not efficient enough[1].

Until an alternative energy source that is cost-effective is found to concentrate on this energy problem, it is critical to make use of the energy that is already accessible and the energy that is stored can be retrieved when it is needed. During sunny days, the heat that is collected can be stored and then used at night in solar energy systems. Buildings in which the requirements for heating are considerable and the cost of electricity enables heat storage to be viable with alternative heating sources can use heat storage. Low melting organic materials are receiving attention for their ability to avoid the difficulties of supercooling and segregation that are associated with inorganic phase change materials[2].

Fossil fuel scarcity means that the demand for energy always exceeds the supply, causing an energy crises. When solar energy is efficiently stored in thermal energy storage systems (TES), the problem is to some extent tolerable without harming the environment. Seasonal changes induce an increase in heating demands, which drives up energy usage. Organic phase change materials (PCMs) are more frequently used than inorganic PCMs for their key qualities, with no issue of supercooling, better thermal properties, and chemical stability. Thermal conductivity of organic PCMs is so high that they are not suitable for energy storage or release[3].

TES systems are critical for overcoming the energy supply-demand mismatch by managing the interval. The application of the Latent Heat Thermal Energy Storage (LTES) system is controlled by the low thermal conductivity of PCMs in many applications. Due to the PCM's low thermal conductivity, it would be more difficult to

DOI: 10.1201/9781003163633-1

1

store energy and the melting and solidification durations would be longer since heat transfer from the heat transfer fluid (HTF) to the PCM would be slower[4, 5].

PCMs were inserted into metal structure heat exchangers to enhance heat transfer performance. It is widely known that metals are better conductors of heat than PCMs in liquid state. Multiple PCMs can also be used to increase energy efficiency in LTES systems rather than using a single PCM. Species with the highest melting points are used in the multiple PCM method. The HTF temperature is kept consistent while charging, even though the flow direction decreases the HTF temperature. While discharging, the HTF flow is directly opposed to charging[6].

1.2 NANOMATERIALS EMBEDDED PHASE CHANGE MATERIALS

In LTES systems, nanofluids with embedded PCMs contain both base PCMs and solid NPs (nanoparticles). Nanofluid is also known as composite PCM since it is composed of two or more disparate materials (PCMs and solid NPs) combined together. When nanofluids fill a spherical or cylindrical encapsulation and are stored in a storage tank, they tend not to flow in the heat exchange systems[7].

NP dispersion in PCM is critical to improving the thermal characteristics of PCM. The Al_2O_3-H_2O nanofluids were found to improve thermal conductivity by 10.5% when 0.2 wt% Al_2O_3 nanoparticles were added to water. Nanoparticles in saturated $BaCl_2$ solution enhanced both the total cool storage and supply by increasing thermal conductivity[8].

The conductivity of single nanofluids (CNTs, Au NPs, and Cu NPs) and hybrid nanofluids (CNT-Au NP and CNT-Cu NP) was examined, as were their two different cooling strategies. The thermal conductivity of nanofluid with Cu NPs improved by 74% over the rest, and the thermal conductivity of Cu NP suspension increased in a linear connection with Cu NP concentration. In order to solidify and melt completely, nanofluids will need additional time. When nanoparticles of Ag are combined into and synthesized into 1-tetradecanol (TD), the concentration of nanoparticles of Ag increases with a sample of PCM with better thermal conductivity[9].

The thermal conductivity of nanofluids has been found to increase as particle size decreases. Brownian motion of nanoparticles promotes this trend and liquid layering accompanies the movement of nanoparticles. The creation of clusters can increase heat conductivity to some extent, but excessive clustering can have the reverse effect due to the associated sedimentation[10].

A building's cooling system relies on PCMs for the storage of cold energy that is removed from the chilled water system during off-peak hours and returned at peak times. The PCMs used in the LTES system would lower the heat transfer rate. For this reason, the PCMs would retain the cold energy for a longer period of time, which would then interpret into increased electrical energy consumption because of the use of a chilled water system. When it comes to heating systems, PCMs may both store and release heat energy according to the time of day or season. Because of the PCM's low thermal conductivity, the heating unit must run for longer periods of time in order to provide the heat energy, which increases the cost of electricity[11].

Solar energy storage is another area of study because it is readily available and can pull in a lot of energy from the solar energy. It's also intermittent, and it relies heavily on the climate conditions. PCMs with high thermal conductivity are preferable for storing solar energy.

1.3 PCMS AND SELECTION

A number of approaches have been developed to improve PCM performance, including the use of metal fins, metal meshes and beads, and impregnation of porous materials. Even though these approaches were only minimally effective, they also created more weight and volume for the LTES systems. Recently, the advancement of nanotechnology has led to the creation of a novel approach, known as "nanofluids," consisting of both base fluid and solid NPs in varying sizes ranging from 1 nm to 100 nm for the purpose of enhancement in thermal conductivity. It has been established that the optimal strategy for thermal conductivity augmentation is dispersion of solid NPs in the base fluid. Metal NPs, metal oxide NPs, graphene, and carbon nanotubes (CNTs) are used to obtain improved thermal conductivity.

As the PCMs are used to store latent heat, the term "latent heat storage materials" is often used to refer to them. Thermal energy is transferred when a substance changes from a solid to a liquid or vice versa. This phenomenon is known as a phase change, or a state process. In the beginning, these PCMs are simply regular storage materials as their temperature increases and they are able to absorb heat. In contradiction to conventional (sensible) storage materials, PCMs' temperature remains relatively constant while absorbing and releasing heat. Their heat-storage capacity is 5 to 14 times that of sensible materials such as water, masonry, and rock.

Many PCMs have been found to melt in the requisite range of heat of fusion although PCMs are anticipated to have certain thermodynamic, kinetic, and chemical features that are desired. Additionally, cost and availability are important factors in considering these materials for usage in LTES systems. PCMs can be divided into three categories: organic, inorganic, and eutectic (as seen in Figure 1.1). The two main categories of organic PCMs are known as paraffins and non-paraffins. Organic PCMs have significant disadvantages: they are significantly more expensive, and have low thermal conductivity.

Salt hydrates are also known as inorganic PCMs. The benefits of salt hydrates include: less cost than organic PCM, higher latent heat per unit weight and volume, and higher thermal conductivity than organic components. The main disadvantage with salt hydrates is supercooling occurrence. A eutectic is a state in which two or more components, each of which melts and freezes, meet to form a mixture of crystals. Both components are liquefied simultaneously, again with an improbable separation.

For cooling applications the PCM melting temperature should be below 15°C with desired latent heat, while it should be above 60°C with a latent heat of 120 kJ/kg and more for heating purposes. With additional heat cycles for PCMs (melting/freezing) to ensure their long-term use, the thermal stability of PCMs should be studied to determine the degradation. In addition, the PCMs should be chosen according to availability and cost. Table 1.1 presents the thermophysical parameters of the PCMs that have been explored for LTES systems.

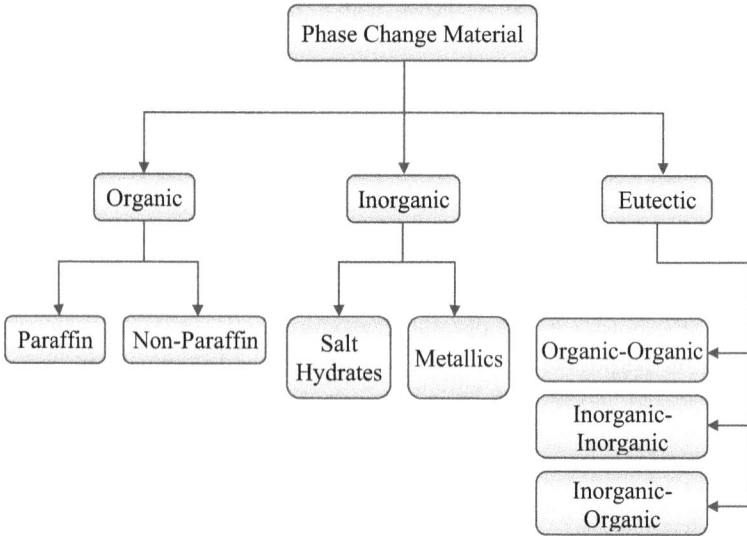

FIGURE 1.1   The categorization of PCMs[12].

TABLE 1.1
Thermophysical Properties of the PCMs Studied forLTES Systems[13]

Compound	Melting Point (°C)	Latent Heat (J/g)	Heat Capacity (J/g·K)	Thermal Conductivity (W/m·K)	Density (g/cm³)
H_2O	0	335	4.2	2.4	1000
n-Octadecane	27.7	243.5	2.66(L)/2.14(S)	1.15(L)/1.19(S)	0.79(L)/0.86(S)
$CaCl_2 \cdot 6H_2O$	29.9	187	2.2(L)/1.4(S)	0.53(L)/1.09(S)	1.53(L)/1.71(S)
Paraffin wax	32.1	251	1.92(L)/3.26(S)	0.224(L)/0.514(S)	0.83
Capric acid	32.0	152.7	-	0.153(L)	0.88(L)/1.00(S)
Polyethylene glycol 900(PEG)	34.0	150.5	2.26	0.188	1.1(L)/1.2(S)
Lauric acid	41–43	211.6	2.27(L)/1.76(S)	1.6	0.86(L)/1.76(S)
Stearic acid	41–43	211.6	2.27(L)/1.76(S)	1.6	0.86(L)/1.00(S)
Paraffin wax	40–53	251	1.92(L)/3.26(S)	0.224(L)/0.514(S)	0.83
Commercial paraffin wax	52.2	243.5	-	0.15	0.77(L)/0.81(S)

1.4   HEAT TRANSFER ENHANCEMENT

The capability of LTES systems to store and release energy has not been thoroughly examined in many of its critical characteristics. PCMs in LTES systems have very low thermal conductivity and significantly reduce heat transfer rates during energy storage and release.

The thermal conductivity of the materials to be improved has been incorporated into the graphite matrix as a promising candidate to encourage the enhancement of heat transfer. The most significant benefit of this material is that it can be used to improve thermal conductivity in PCMs without reducing energy storage, but the material also offers a reduction in salt hydrate sub cooling and paraffin volume. The styrene-butadiene-styrene triblock copolymer and exfoliated graphite was produced based on paraffin. In experiments, it was discovered that no paraffin leakage was present in the phase change process, and the material had excellent thermal conductivity, along with about 80% of the latent heat of fusion per unit weight of paraffin[14, 15].

REFERENCES

[1] Castell, A., Martorell, I., Medrano, M., Perez, G. & Cabeza, L. F., 'Experimental study of using PCM in brick constructive solutions for passive cooling', *Energy and Buildings*, vol. 42, pp. 534–540, 2010.

[2] Dolado, P., Lazaro, A., Marin, J. M. & Zalba, B., 'Characterization of melting and solidification in a real-scale PCM-air heat exchanger: Experimental results and empirical model', *Renewable Energy*, vol. 36, pp. 2906–2917, 2011.

[3] Fan, L.-W., Xiao, Y.-Q., Zeng, Y., Fang, X., Wang, X., Xuc, X., Yu, Z.-T., Rong-Hua Hong, R.-H., Hua, Y.-C. & Cen, K.-F., 'Effects of melting temperature and the presence of internal fins on the performance of a phase change material (PCM)-based heat sink', *International Journal of Thermal Sciences*, vol. 70, pp. 114–126, 2013.

[4] Leong, K. Y., Rahman, M. R. A. & Gurunathan, B. A., 'Nano-enhanced phase change materials: A review of thermo-physical properties, applications and challenges', *Journal of Energy Storage*, vol. 21, pp. 18–31, 2019.

[5] Khalifa, A., Tan, L., Date, A. & Akbarzadeh, A., 'A numerical and experimental study of solidification around axially finned heat pipes for high temperature latent heat thermal energy storage units', *Applied Thermal Engineering*, vol. 70, no. 1, pp. 609–619, 2014.

[6] Kazemi, I., Sefid, M. & Afrand, M., 'Improving the thermal conductivity of water by adding mono & hybrid nano-additives containing graphene and silica: A comparative experimental study', *International Communications in Heat and Mass Transfer*, vol. 116, p. 104648, 2020.

[7] Chiam, H. W., Azmi, W. H., Usri, N. A., Mamat, R. & Adam, N. M., 'Thermal conductivity and viscosity of Al2O3 nanofluids for different based ratio of water and ethylene glycol mixture', *Experimental Thermal and Fluid Science*, vol. 81, pp. 420–429, 2017.

[8] Harikrishnan, S., Imran Hussain, S., Devaraju, A., Sivasamy, P. & Kalaiselvam, S., 'Improved performance of a newly prepared nano-enhanced phase change material for solar energy storage', *Journal of Mechanical Science and Technology*, vol. 31, no. 10, pp. 4903–4910, 2017.

[9] Karaipekli, A. & Sarı, A., 'Capric: Myristic acid/expanded perlite composite as form-stable phase change material for latent heat thermal energy storage', *Renewable Energy*, vol. 33, no. 12, pp. 2599–2605, 2008.

[10] Oró, E., De Gracia, A., Castell, A., Farid, M. M. & Cabeza, L. F., 'Review on phase change materials (PCMs) for cold thermal energy storage applications', *Applied Energy*, vol. 99, pp. 513–533, 2012.

[11] Wang, B., Wang, X., Lou, W. & Hao, J., 'Ionic liquid-based stable nanofluids containing gold nanoparticles', *Journal of Colloid and Interface Science*, vol. 362, pp. 5–14, 2011.

[12] Wu, S., Zhu, D., Li, X., Li, H. & Lei, J., 'Thermal energy storage behavior of Al_2O_3_H_2O nanofluids', *Thermochimica Acta*, vol. 483, pp. 73–77, 2009.

[13] Xuan, Y. & Li, Q., 'Heat transfer enhancement of nanofluids', *International Journal of Heat and Fluid Flow*, vol. 21, pp. 58–64, 2000.

[14] Zalba, B., Marin, J. M., Cabeza, L. F. & Mehling, H., 'Free-cooling of buildings with phase change materials', *International Journal of Refrigeration*, vol. 27, pp. 839–849, 2004.

[15] Zeng, J. L., Cao, Z., Yang, D. W., Xu, F., Sun, L. X., Zhang, X. F. & Zhang, L., 'Effects of MWNTs on phase change enthalpy and thermal conductivity of a solid-liquid organic PCM', *Journal of Thermal Analysis and Calorimetry*, vol. 2, pp. 507–512, 2009.

2 NanoPCMs

CONTENTS

2.1 NANOFLUIDS AS PCMS

The thermal conductivity of base fluids has not been able to be increased to the appropriate level in recent investigations. Nanotechnology has introduced a new technique, "nanofluids," that includes solid NPs and base fluids to improve the system's performance[1]. The particles of nanometer size have a large potential to improve the heat transfer of the base liquids. Nanofluids have greater advantages than conventional solid-liquid suspensions for enhancing heat transfer and are as follows[2–4]:

1. There is better heat transfer between particles and liquids because of the higher specific surface area.
2. The dispersion stability is high, and there is dominant Brownian motion.
3. Heat transfer is more efficient in comparison with pure liquid, and reduction in pumping power.
4. Miniaturization of the system is encouraged by having a reduced particle block than conventional slurries.
5. Flexible properties, including thermal conductivity and surface wettability, can suit a variety of applications with various particle concentrations.

The thermal conductivity of base fluids is increased with the preference of metal NPs (Cu, Al, Ag, etc.), metal oxide NPs (CuO, Al$_2$O$_3$, ZnO, etc.), Graphens, and CNTs. Preparation of nanofluids serves an essential task in improving the thermal and heat transfer effects of nanofluids. The productive nanofluid preparation process uses various NP synthesis approaches. In early times nanofluids were used as oxide particles, which were easy to create and improved chemical stability[5].

During nanofluid preparation, homogenous dispersion and longer NP stability, together with little or no agglomeration of the NPs in the base fluids or base materials, should be taken into consideration[6, 7]. Nanofluid PCM is created by dispersing or suspending the entire amount of NPs in the melted PCM, and hence it is also known as composite PCM. Increased thermal conductivity and reduced supercooling is achieved by adding NPs into the PCMs. In addition, insertion of NPs in the PCMs can modify phase temperatures and latent heat from the base values. Thus, it is vital

DOI: 10.1201/9781003163633-2

to determine that these phase changes should be as slight as possible and have no impact on the LTES performance and heat storage capacity[8].

The improvement would be more effective for spherical NPs, since the area for the given volume is higher in comparison to the cylindrical, square, rectangular, and irregular shapes. It explored the appropriate volume fraction of NPs needed for more rapid PCM solidification process. In the next case, particle aggregation increases and base fluid thermal conductivity decreases, both of which can result in an agglomeration of the base fluid NPs[9].

2.2   HEAT CAPACITY

A nanofluid's heat capacity has a propensity to be smaller than that of the base fluid, and it varies according to the volume or mass fraction of NPs distributed in the base fluid. As the mass fraction of the NP in nanofluids increases, nanofluids lower their heat capacity. It shows that heat energy required for nanofluids would be less for the same temperature increase compared to base fluid. Paraffin-based nanofluid phase change behavior with Cu NPs with different mass fractions is examined. The results of the Differential Scanning Calorimetry (DSC) predict a maximum reduction of 11.1% and 11.7% in melting latent heat and freezing latent heat[10–15].

A systematic experimental research was performed using the DSC to examine the specific heat of a water-based nanofluid that contained Al_2O_3. As the NPs' mass fraction increases, the nanofluid's specific heat steadily drops. To confirm these results, in another study, the researchers found that adding carbon nanofillers decreases the energy storage capacity. Thus, the DSC test was conducted to detect the impacts of different carbon nanofillers on the melting/solidification enthalpy of the composite PCM[16].

In another work, nanofluid for low-temperature cool storage systems were created with aqueous $BaCl_2$ solution integrated with TiO_2 NPs. The results indicated that the latent heat of nanofluids at 0.167% of volume decreases very little, but latent heat decreases dramatically with a volume fraction of 0.283%.

It has been concluded that the concentration of $BaCl_2$ aqueous solution dropped significantly by 0.283 vol% NPs. On the contrary, the average increase in the specific heat capacity was 14.5%. This shows that there are alternative transfer processes for nanofluids, which distinguishes nanofluids from a mix of conventional fluids[17].

2.3   THERMAL CONDUCTIVITY

Thermal conductivity is the ability of a material to conduct or transfer heat. A thermal conductivity of materials is regarded as one of the primary heat transfer rates. The maximum thermal conductivity improved for nanofluids was 32.4% for a volume fraction of 4.3%. It was also discovered that the thermal conductivity of the material was directly proportional to the volume fraction of the particles. An experimental test was conducted on nanofluids that contain Al_2O_3 and CuO in water and ethylene glycol to estimate thermal conductivity. The thermal conductivity is shown to be linearly related to particle volume fraction. For 4% of CuO NP's volume distributed into ethylene glycol, the greatest increase in thermal conductivity was recorded as 20%. Improvement of the thermal conductivity to 40% has been achieved for 8% volume fraction of Al_2O_3 NPs[18–24].

The effect in water and ethylene glycol of Al_2Cu and Ag_2Al NPs is investigated. Thermal conductivity of nanofluids was studied at room temperature, and the results show that the thermal conductivity of the Ag_2Al NPs was higher than the Al_2Cu NPs. It was carried out by comparing the double-walled CNT (DWCNT) and multiwall CNT nanofluids (MWCNT). The maximum thermal conductivity improvement of 34% was attained for the 0.6% of the volume portion of MWCNT, but a 3% increase was achieved for the 0.75% volume fraction of DWCNT. The reason for this low improvement is that the DWCNT has been transformed into micrometers by cluster effects[25–27].

The thermal conductivity of a combination made with ethylene glycol and synthetic engine oil was discovered by mixing it with MWCNT. Thermal conductivity of the nanofluid was found to be 12.4% when using 1% MWCNT/ethylene glycol nanofluid, whereas thermal conductivity was found to be 22% when using 2% MWCNT/ethylene glycol nanofluid. The MWCNT/synthetic engine oil nanofluid displayed thermal conductivity of 12.4%[28–30].

Nanofluid thermal conductivity is proportional to the NP's diameter. A variety of NPs, including those as low as 1 nm to as large as 100 nm, are made through a variety of synthetic processes. In the investigation of nanofluids, ethylene glycol and copper nanoparticles (NPs) were examined. Thermal conductivity can be increased by 40% by using Cu NPs smaller than 10 nm with a concentration of 0.3%. The thermal conductivity of nanofluids consisting of Al_2O_3 NPs and water as the base fluid was investigated. Thermal conductivity enhancement might be investigated by comparing the 36 nm and 47 nm Al_2O_3 NPs. A nanofluid with a particle size of 36 nm was created using Al_2O_3 and water, and it was shown to have the maximum effect when heated to higher temperatures[31–35].

The impact of varying NP shape on improved thermal conductivity is investigated, as each shape has a different surface structure and so potentially provides different heat transfer contact surfaces. Spherical and cylindrical shape of the nanofluids was separately synthesized. In nanofluids containing cylindrical NPs, higher thermal conductivity improvements were seen as compared with nanofluids containing spherical NPs[36].

Experimental studies on nanofluids at temperatures of 13, 23, 40, and 55°C were done to estimate thermal conductivity, with varying volume fractions of TiO_2 NPs. The thermal conductivity ratio was not considerably affected by the fluctuating temperature. Temperature effect on nanofluid thermal conductivity has been examined. Based on the results, the thermal conductivity ratio was increased to temperature at a constant particle volume fraction. Table 2.1 shows the thermal conductivity of several of the nanofluids that were studied utilizing the metal and metal oxide NPs[46–48].

TABLE 2.1

Thermal Conductivity Enhancement of Some of the Nanofluids Using Metal and Metal Oxide NPs[37–45]

Particle	Base Fluid	Average Size of Particle	Volume Fraction	Thermal Conductivity Enhancement
TiO_2	Water	15 nm	5%	30%
Al_2O_3	Water	20 nm	0.2%	9.5%
CuO	Water	50 nm	0.4%	17%
CuO	Ethylene glycol	25 nm	5%	22%
Al_2O_3	Water	10 nm	0.5%	100%
TiO_2	$BaCl_2$-H_2O	20 nm	1.130%	16.74%
Cu	Paraffin	25 nm	1%	10.17%
Ag	Water	60–80 nm	0.001%	17%
Fe	Ethylene glycol	10 nm	0.55%	18%
Cu	Water	100 nm	7.5%	78%
Au	Toluene	15 nm	0.011%	8.8%
SiC	Water	26 nm	4.2%	16%

2.4 VISCOSITY

Using nanofluids in a system that features an adverse viscosity can decrease the system's heat transfer capability. As the fluid's viscosity increases, the flow of the fluid is slowed, and increases the pumping power requirements. The effective viscosity of the Al_2O_3/water nanofluids was reported for different NP concentrations. It shows that linear response can be achieved up to 2% volume fraction, while nonlinear response can be recorded higher than this concentration. This occurs due to the hydrodynamic interactions between the NPs. This is crucial because of the collective effect of fluctuations in the fluid that arise as disturbances around one NP interacting with those around other NPs[49–51].

The effects of volume fraction and temperature on nanofluid viscosity produced by means of TiO_2 NPs and water were examined. The rheological properties of CuO NPs were explored in a 60:40 (by weight) ethylene glycol/water mixture. In this analysis, Newtonian behavior was found for the nanofluid particle volume fraction. At −35°C, the volume fraction of 6.12% nanofluid is approximately four times as large as the base fluid's viscosity. Furthermore, the nanofluids' viscosity decreases exponentially as their temperature increases. The influence of temperature and particulate content in Al_2O_3/water nanofluids on dynamic viscosity has been investigated. Viscosity changes are observed in nanofluids when particle concentration is high, although viscosity decreases as temperature augments[52–58].

An investigation was carried out to measure viscosity for Al_2O_3/decane and Al_2O_3/ isoparaffin polyalphaolefin. Nanofluids without aggregate particles were shown in the protensive viscosity measurements. The experimental research of the viscosity on ethylene glycol (EG) based Al_2O_3 nanofluids was also tested. This conjecture is made on the relative enhancement of the viscosity of the Al_2O_3 nanofluids to the water-based nanofluids. They found that pH influences viscosity as well[59-68].

The viscosity of the nanofluids that contained α-Fe_2O_3 NPs and glycerol was investigated. The experiments showed that, when the amount of the particles in the solution increases, the viscosity of α-Fe_2O_3/glycerol nanofluids increases and lowers in increasing temperature. The fluids behaved non-Newtonian at lower temperatures[69]. Also, because of the presence of nanoparticles in the base fluid, the nanofluids have a relent stress. Base fluid viscosity has a significant impact on the thermal conductivity of nanofluids distributed with Fe_3O_4 NPs[70-74].

In nanofluids, the thermal conductivity decreases with increase of viscosity, but the thermal conductivity reaches a constant value. Increased thermal conductivity can be achieved by lowering the viscosity value of a nanofluid. The viscosity of the nanofluids was measured by placing SiC NPs in a combination of EG/H_2O with a concentration of 50:50 (by volume). The viscosity of a liquid reduces as the average particle size increases at the same temperature. In situations where heat transfer is important, this reduced viscosity is more helpful, reducing the pumping power. According to the result, the viscosity enhancement was found to be unaffected by the viscosity of the base fluid, and instead is affected solely by the particle volume fraction[74-79].

In nanofluids with volume fractions under 5%, Newtonian liquid behavior was seen; however, at volume fractions of 5% or greater, shear shining behavior was evident. The distinction between Newtonian and non-Newtonian behavior of nanofluids is based mostly on the types and shapes of NPs, their volume fractions, and temperature. The viscosity of nanofluids is measured when it contains ultra-dispersed diamond (UDD)/ethylene glycol, silver/water, and silica/water. At volume concentrations of 1%, 2%, and 3%, the viscosity of UDD/EG nanofluid increased by 50%, silver/water nanofluid increased by 30%, and silica/water nanofluid increased by 20%[80-82].

The viscosity of three types of nanofluids was tested experimentally, like Al_2O_3/water, Al_2O_3/EG, and CuO/EG nanofluids. Newtonian behavior was seen in both water-based and ethylene glycol-based nanofluids. The rheological behavior of agglomerated silver NPs distributed into diethylene glycol has been investigated. The major dependence of relative viscosity on NP concentration has indicated that the particulate interactions are increasingly evident with the increasing NP concentration. The critical load value for NPs has been determined at approximately 11%[83].

To evaluate the performance of dispersion methods, they measured the nanofluids' particle size distribution and time stability. It was found that particle size has a large effect on viscosity, and should be factored in when using practical applications. A further study evaluated CuO NPs mixed with water, with weight fractions ranging from 0.05% to 10%. The fluid's viscosity is increased because of the size influence on the electric double layer repulsion caused by NPs with smaller particle size[84-89].

Experimental investigation demonstrated that nanofluid (with Al_2O_3 NPs dispersed in it) has the same viscosity as commercial automotive coolant. It was reported that at room temperature, the pure base fluid acts like a Newtonian fluid, but as soon as an Al_2O_3 NPs concentration is added, it starts behaving like a non-Newtonian fluid. The viscosity of the nanofluid also increases as the amount of NPs in the solution increases, and it decreases as the temperature increases. Nanofluids contain viscosity-enhancing NPs and feature a size and shape dependency of viscosity enhancement. It is believed that nanofluid viscosity enhancement could be divided into three primary parts: NP size, shape, and volume fraction of the NPs, and nanofluid temperature[90, 91].

REFERENCES

[1] Abareshi, M., Sajjadi, S. H., Zebarjad, S. M. & Goharshadi, E. K., 'Fabrication, characterization, and measurement of viscosity of α-Fe₂O₃-glycerol nanofluids', *Journal of Molecular Liquids*, vol. 163, pp. 27–32, 2011.

[2] Abhat, A., Aboul-Enein, S. & Malatidis, N., 'Heat of fusion storage systems for solar heating applications', In: *Thermal Storage of Solar Energy*, Dordrecht: Springer, pp. 157–171, 1981.

[3] Adine, H. A. & El Qarnia, H., 'Numerical analysis of the thermal behaviour of a shell-and-tube heat storage unit using phase change materials', *Applied Mathematical Modelling*, vol. 33, pp. 2132, 2009.

[4] Akoh, H., Tsukasaki, Y., Yatsuya, S. & Tasaki, A., 'Magnetic properties of ferromagnetic ultrafine particles prepared by vacuum evaporation on running oil substrate', *Journal of Crystal Growth*, vol. 45, pp. 495–500, 1978.

[5] Esfe, M. H., Alidoust, S. & Esmaily, R., 'A comparative study of rheological behavior in hybrid nano-lubricants (HNLs) with the same composition/nanoparticle ratio characteristics and different base oils to select the most suitable lubricant in industrial applications', *Colloids and Surfaces A: Physicochemical and Engineering Aspects*, p. 128658, 2022.

[6] Assael, M. J., Metaxa, I. N., Arvanitidis, J., Christofilos, D. & Lioutas, C., 'Thermal conductivity enhancement in aqueous suspensions of carbon multi-walled and double-walled nanotubes in the presence of two different dispersants', *International Journal of Thermophysics*, vol. 26, no. 3, pp. 647–664, 2005.

[7] Beck, M. P., Yuan, Y., Warrier, P. & Teja, A. S., 'The effect of particle size on the thermal conductivity of alumina nanofluids', *Journal of Nanoparticle Research*, vol. 11, no. 5, pp. 1129–1136, 2009.

[8] Cabeza, L. F., Mehling, H., Hiebler, S. & Ziegler, F., 'Heat transfer enhancementin water when used as PCM in thermal energy storage', *Applied Thermal Engineering*, vol. 22, pp. 1141–1151, 2002.

[9] Chandrasekar, M., Suresh, S. & Chandra Bose, A., 'Experimental investigations and theoretical determination of thermal conductivity and viscosity of Al₂O₃/water nanofluid', *Experimental Thermal and Fluid Science*, vol. 34, pp. 210–216, 2010.

[10] Choi, S. U. S., Zhang, Z. G., Yu, W., Lockwood, F. E. & Grulke, E. A., 'Anomalous thermal conductivity enhancement in nanotube suspensions', *Applied Physics Letter*, vol. 79, no. 14, pp. 2252–2254, 2001.

[11] Choi, S.U.S., In: Singer, D. A. & Wang, H. P. (Eds.), *Development and Application of Non-Newtonian Flows*, vol. FED 231, New York: ASME, pp. 99–105, 1995.

[12] Chon, C. H., Kihm, K. D., Lee, S. P. & Choi, S. U. S., 'Empirical correlation finding the role of temperature and particle size for nanofluid (Al₂O₃) thermal conductivity enhancement', *Applied Physics Letter*, vol. 87, no. 15, pp. 153107, 2005.

[13] Chon, C. H. & Kihm, K. D., 'Thermal conductivity enhancement of nanofluids by Brownian motion', *Journal of Heat Transfer*, vol. 127, no. 8, p. 810, 2005.

[14] Chopkar, M., Sudarshan, S., Das, P. K. & Manna, I., 'Effect of particle size on thermal conductivity of nanofluid', *Metallurgical Transactions A, Physical Metallurgy and Materials Science*, vol. 39, no. 7, pp. 1535–1542, 2008.

[15] Chopkar, M., Das, P. K. & Manna, I., 'Synthesis and characterization of nanofluid for advanced heat transfer applications', *Scripta Materialia*, vol. 55, no. 6, pp. 549–552, 2006.

[16] Costa, M., Buddhi, D. & Oliva, A., 'Numerical simulation of a latent heat thermalenergy storage system with enhanced heat conduction', *Energy Conversion and Management*, vol. 39, no. 3/4, pp. 319–330, 1998.

[17] Czarnetzki, W. & Roetzel, W., 'Temperature oscillation techniques for simultaneous measurement of thermal diffusivity and conductivity', *International Journal of Thermophysics*, vol. 16, no. 2, pp. 413–422, 1995.

[18] Das, S. K., Putra, N., Thiesen, P. & Roetzel, W., 'Temperature dependence of thermal conductivity enhancement for nanofluids', *Journal of Heat Transfer*, vol. 125, no. 4, pp. 567–574, 2003.

[19] Ding, Y., Alias, H., Wen, D. & Williams, R. A., 'Heat transfer of aqueous suspensions of carbon nanotubes (CNT nanofluids)', *International Journal of Heat and Mass Transfer*, vol. 49, nos. 1–2, pp. 240–250, 2006.

[20] Duangthongsuk, W. & Wongwises, S., 'Measurement of temperature-dependent thermal conductivity and viscosity of TiO$_2$-water nanofluids', *Experimental Thermal and Fluid Science*, vol. 33, pp. 706–714, 2009.

[21] Eastman, J. A., Choi, S. U. S., Li, S. & Thompson, L. J., 'Enhanced thermal conductivity through the development of nanofluids', In: *Proceedings of the Symposium on Nanophase and Nanocomposite Materials II*, vol. 457, Materials Research Society, USA, pp. 3–11, 1997.

[22] Elgafy, A. & Lafdi, K., 'Effect of carbon nanofiber additives on thermal behavior of phase change materials', *Carbon*, vol. 43, pp. 3067–3074, 2005.

[23] Esen, M. & Durmus, A., 'Geometric design of solar-aided latent heat store depending on various parameters and phase change materials', *Solar Energy*, vol. 62, pp. 19, 1998.

[24] Fan, L.-W., Fang, X., Wang, X., Zeng, Y., Xiao, Y.-Q., Yu, Z.-T., Xu, X., Hu, Y.-C. & Cen, K.-F., 'Effects of various carbon nanofillers on the thermal conductivity and energy storage properties of paraffin-based nanocomposite phase change materials', *Applied Energy*, vol. 110, pp. 163–172, 2013.

[25] Fukai, J., Hamada, Y., Morozumi, Y. & Miyatake, O., 'Improvement of thermal characteristics of latent heat thermal energy storage units using carbon-fiber brushes: Experiments and modeling', *International Journal of Heat and Mass Transfer*, vol. 46, pp. 4513.172, 2003.

[26] George, A., 'Hand book of thermal design', In: Guyer, C. (Ed.), *Phase Change Thermal Storage Materials*. McGraw Hill Book Co. [chapter 1], 1989.

[27] Gleiter, M., 'Nanocrystalline materials', *Progress in Materials Science*, vol. 33, pp. 223–315, 1989.

[28] Haddad, Z., Abid, C., Oztop, H. F. & Mataoui, A., 'A review on how the researchers prepare their nanofluids', *International Journal of Thermal Sciences*, vol. 76, pp. 168–189, 2014.

[29] Hamilton, R. L. & Crosser, O. K., 'Thermal conductivity of heterogeneous two-component systems', *Industrial and Engineering Chemistry Fundamentals*, vol. 1, no. 3, pp. 187–191, 1962.

[30] Moradikazerouni, A., 'Heat transfer characteristics of thermal energy storage system using single and multi-phase cooled heat sinks: A review', *Journal of Energy Storage*, vol. 49, p. 104097, 2022.

[31] He, Q., Wang, S., Tang, M. & Liu, Y., 'Experimental study on thermophysical properties of nanofluids as phase-change material (PCM) in low temperature cool storage', *Energy Conversion and Management*, vol. 64, pp. 199–205, 2012.

[32] Hong, T.-K., Yang, H.-S. & Choi, C. J., 'Study of the enhanced thermal conductivity of Fe nanofluids', *Journal of Applied Physics*, vol. 97, no. 6, pp. 1–4, 2005.

[33] Huang, M. J., Eames, P. C. & Norton, B., 'Thermal regulation of building-integrated photovoltaics using phase change materials', *International Journal of Heat and Mass Transfer*, vol. 47, p. 2715, 2004.

[34] Hwang, Y. J., Ahn, Y. C., Shin, H. S., Lee, C. G., Kim, G. T., Park, H. S. & Lee, J. K., 'Investigation on characteristics of thermal conductivity enhancement of nanofluids', *Current Applied Physics*, vol. 6, pp. e67–71, 2006.

[35] Qin, M., Almohsen, B., Sabershahraki, M. & Issakhov, A., 'Investigation of water freezing with inclusion of nanoparticle within a container with fins', *Applied Nanoscience*, pp. 1–13, 2022.

[36] Ju, Y. S., Kim, J. & Hung, M. T., 'Experimental study of heat conduction in aqueous suspensions of aluminum oxide nanoparticles', *Journal of Heat Transfer*, vol. 130, no. 9, p. 092403, 2008.

[37] Kalaiselvam, S., Parameshwaran, R. & Harikrishnan, S., 'Analytical and experimental investigations of nanoparticles embedded phase change materials for cooling application in modern buildings', *Renewable Energy*, vol. 39, pp. 375–387, 2012.

[38] Kang, H. U., Kim, S. H. & Oh, J. M., 'Estimation of thermal conductivity of nanofluidusing experimental effective particle volume', *Experimental Heat Transfer*, vol. 19, pp. 181–191, 2006.

[39] Kenisarin, M. & Mahkamov, K., 'Solar energy storage using phase change materials', *Renewable and Sustainable Energy Reviews*, vol. 11, pp. 1913–1965, 2007.

[40] Goodarzi, M., Esfandeh, S. & Toghraie, D. 'A state of art review of the viscosity behavior of nano-lubricants containing MWCNT nanoparticles: Focusing on engine lubrication goals', *Journal of Molecular Liquids*, vol. 346, p. 118264, 2022.

[41] Kulkarni, D. P., Das, D. K. & Vajjha, R. S., 'Application of nanofluids in heating buildings and reducing pollution', *Applied Energy*, vol. 86, no. 12, pp. 2566–2573, 2009.

[42] Lane, G. A., 'Solar Heat Storage: Latent Heat Materials', *Background and Scientific Principles*, vol. 1, Florida: CRC Press, Inc., 1983.

[43] Lee, S., Choi, S. U. S., Li, S. & Eastman, J. A., 'Measuring thermal conductivity of fluids containing oxide nanoparticles', *Journal of Heat Transfer*, vol. 121, pp. 280–289, 1999.

[44] Li, C. H. & Peterson, G. P., 'Experimental investigation of temperature and volume fraction variations on the effective thermal conductivity of nanoparticle suspensions (nanofluids)', *Journal of Applied Physics*, vol. 99, no. 8, pp. 1–8, 2006.

[45] Liu, M.-S., Lin, M. C.-C., Huang, I.-T. & Wang, C.-C., 'Enhancement of thermal conductivity with carbon nanotube for nanofluids', *International Communications in Heat and Mass Transfer*, vol. 32, no. 9, pp. 1202–1210, 2005.

[46] Liu, Y. D., Zhou, Y. G., Tong, M. W. & Zhou, X. S., 'Experimental study of thermal conductivity and phase change performance of nanofluids PCMs', *Microfluidics and Nanofluidics*, vol. 7, no. 4, pp. 579–584, 2009.

[47] Lo, C.-H., Tsung, T.-T. & Chen, L.-C., 'Shape-controlled synthesis of Cu based nanofluid using submerged arc nanoparticle synthesis system (SANSS)', *Journal of Crystal Growth*, vol. 277, no. 1–4, pp. 636–642, 2005.

[48] Masuda, H., Ebata, A., Teramae, K. & Hishinuma, N., 'Alteration of thermal conductivity and viscosity of liquid by dispersing ultrafine particles (dispersion of c-Al_2O_3, SiO_2, and TiO_2 ultra-fine particles)', *Netsu Bussei*, vol. 4, no. 4, pp. 227–233, 1993.

[49] Medrano, M., Yilmaz, M. O., Nogues, M., Martorell, I., Roca, J. & Cabeza, L. F., 'Experimental evaluation of commercial heat exchangers for use as PCM thermal storage systems', *Applied Energy*, vol. 86, pp. 2047, 2009.

[50] Mehling, H., Hiebler, S. & Ziegler, F., 'Latent heat storage using a PCM-graphite composite material', In: *Proceedings of Terrastock 2000–8th International Conference on Thermal Energy Storage*, Stuttgart, Germany, pp. 375–380, 2000.

[51] Mintsa, H. A., Roy, G., Nguyen, C. T. & Doucet, D., 'New temperature dependent thermal conductivity data for water-based nanofluids', *International Journal of Thermal Sciences*, vol. 48, no. 2, pp. 363–371, 2009.

[52] Mosaffa, A. H., Infante Ferreira, C. A., Rosen, M. A. & Talati, F., 'Thermal performance optimization of free cooling systems using enhanced latent heat thermal storage unit', *Applied Thermal Engineering*, vol. 59, pp. 473–479, 2013.

[53] Murshed, S. M. S., Leong, K. C. & Yang, C., 'Investigations of thermal conductivity and viscosity of nanofluids', *International Journal of Thermal Sciences*, vol. 47, no. 5, pp. 560–568, 2008b.

[54] Murshed, S. M. S., Leong, K. C. & Yang, C., 'Enhanced thermal conductivity of TiO_2—water based nanofluids', *International Journal of ThermalSciences*, vol. 44, pp. 367–373, 2005.

[55] Namburu, P. K., Kulkarni, D. P., Misra, D. & Das, D. K., 'Viscosity of copper oxide nanoparticles dispersed in ethylene glycol and water mixture', *Experimental Thermal and Fluid Science*, vol. 32, pp. 397–402, 2007.

[56] Nguyen, C. T., Desgranges, F., Galanis, N., Roy, G., Mare, T. & Boucher, S., et al., 'Viscosity data for Al_2O_3/water nanofluid-hysteresis: Is heat transfer enhancement using nanofluids reliable?' *International Journal of Thermal Science*, vol. 47, pp. 103–111, 2008.

[57] Ozerinc, S., Kakac, S. & Yazicioglu, A. G., 'Enhanced thermal conductivity of nanofluids: A state-of-the-art review', *Microfluidics and Nanofluidics*, vol. 8, pp. 145–170, 2010.

[58] Zhao, C., Opolot, M., Liu, M., Wang, J., Bruno, F., Mancin, S. & Hooman, K., 'Review of analytical studies of melting rate enhancement with fin and/or foam inserts', *Applied Thermal Engineering*, p. 118154, 2022.

[59] Pastoriza-Gallego, M. J., Casanova, C., Páramo, R., Barbés, B., Legido, J. L. & Pineiro, M. M., 'A study on stability and thermophysical properties (density and viscosity) of Al_2O_3 in water nanofluid', *Journal of Applied Physics*, vol. 106, pp. 064301, 2009.

[60] Patel, H. E., Das, S. K. & Sundararajan, T., 'Thermal conductivities of naked and monolayer protected metal nanoparticle based nanofluids: Manifestation of anomalous enhancement and chemical effects', *Applied Physics Letter*, vol. 83, pp. 2931–2933, 2003.

[61] Putnam, S. A., Cahill, D. G., Braun, P. V., Ge, Z. & Shimmin, R. G., 'Thermal conductivity of nanoparticles suspensions', *Journal of Applied Physics*, vol. 99, no. 8, pp. 084308, 2006.

[62] Putnam, S. A., Cahill, D. G., Braun, P. V., Ge, Z. & Shimmin, R. G., 'Thermal conductivity of nanoparticles suspensions', *Journal of Applied Physics*, vol. 99, no. 8, pp. 084308, 2006.

[63] Py, X., Olives, R. & Mauran, S., 'Paraffin/porous-graphite-matrix composite as a high and constant power thermal storage material', *International Journal of Heat and Mass Transfer*, vol. 44, pp. 2727–2737, 2001.

[64] Roy, G., Nguyen, C. T., Doucet, D., Suiro, S. & Mare, T., 'Temperature dependent thermal conductivity of alumina based nanofluids', In: Davis, G. V. & Leonardi, E. (Eds.), *Proceedings of 13th International Heat Transfer Conference*, Redding, CT: Begell House Inc, 2006.

[65] Schmidt, J., Chiesa, M., Torchinsky, D. H., Johnson, J. A., Boustani, A. & McKinley, G. H., et al., 'Experimental investigation of nanofluid shear and longitudinal viscosities', *Applied Physics Letters*, vol. 92, pp. 244107, 2008.

[66] Sharma, A., Tyagi, V. V., Chen, C. R. & Buddhi, D., 'Review on thermal energy storage with phase change materials and applications', *Renewable and Sustainable Energy Reviews*, vol. 13, pp. 318–345, 2009.

[67] Sharma, A., Won, L. D., Buddhi, D. & Park, J. U., 'Numerical heat transfer studies of the fatty acids for different heat exchanger materials on the performance of a latent heat storage system', *Renewable Energy*, vol. 30, pp. 2179, 2005.

[68] Shin, D. & Banerjee, D., 'Enhancement of specific heat capacity of high temperature silica-nanofluids synthesized in alkali chloride salt eutectics for solar thermal-energy storage applications', *International Journal of Heat and Mass Transfer*, vol. 54, pp. 1064–1070, 2011.

[69] Siegel, R., 'Solidification of low conductivity material containing dispersed high conductivity particles', *International Journal of Heat and Mass Transfer*, vol. 20, pp. 1087–1089, 1977.

[70] Tamjid, E. & Guenther, B. H., 'Rheology and colloidal structure of silver nanoparticles dispersed in diethylene glycol', *Powder Technology*, vol. 197, pp. 49–53, 2010.

[71] Timofeeva, E. V., Routbort, J. L. & Singh, D., 'Particle shape effects on thermophysical properties of alumina nanofluids', *Journal of Applied Physics*, vol. 106, no. 1, pp. 014304, 2009.

[72] Timofeeva, E. V., Yu, W., France, D. M., Singh, D. & Routbort, J. L., 'Base fluid and temperature effects on the heat transfer characteristics of SiC in ethylene glycol/H_2O and H_2O nanofluids', *Journal of Applied Physics*, vol. 109, pp. 014914, 2011.

[73] Tsai, T. H., Kuo, L. S., Chen, P. H. & Yang, C. T., 'Effect of viscosity of base fluid onthermal conductivity of nanofluids', *Applied Physics Letters*, vol. 93, no. 23, pp. 233121, 2008.

[74] Tuncbilek, K., Sari, A., Tarhan, S., Ergunes, G. & Kaygusuz, K., 'Lauric and palmitic acids eutectic mixture as latent heat storage material for low temperature heating applications', *Energy*, vol. 30, pp. 677, 2005.

[75] Turgut, A., Tavman, I., Chirtoc, M., Schuchmann, H. P., Sauter, C. & Tavman, S., 'Thermal conductivity and viscosity measurements of water-based TiO2 nanofluids', *International Journal of Thermophysics*, vol. 1–14, 2009, doi:10.1007/s10765-009-0594-2.

[76] Velraj, R., Seeniraj, R. V., Hafner, B., Faber, C. & Schwarzer, K., 'Heat transfer enhancement in a latent heat storage system', *Solar Energy*, vol. 65, pp. 171–180, 1999.

[77] Wang, Y., Amiri, A. & Vafai, K., 'An experimental investigation of the melting process in a rectangular enclosure', *Heat and Mass Transfer*, vol. 42, pp. 3659, 1999.

[78] Rasool, M. H., Zamir, A., Elraies, K. A., Ahmad, M., Ayoub, M. & Abbas, M. A., 'A deep eutectic solvent based novel drilling mud with modified rheology for hydrates inhibition in deep water drilling', *Journal of Petroleum Science and Engineering*, 110151, 2022.

[79] Wang, X., Xu, X. & Choi, S. U. S., 'Thermal conductivity of nanoparticle: Fluid mixture', *Journal of Thermophysics and Heat Transfer*, vol. 13, no. 4, pp. 474–480, 1999.

[80] Wagener, M., Murty, B. S. & Gunther, B., 'Preparation of metal nanosuspensions by high-pressure DC-sputtering on running liquids', In: Komarnenl, S., Parker, J.C. & Wollenberger, H. J. (Eds.), *Nanocrystalline and Nanocomposite Materials II*, vol. 457, Pittsburgh, PA: Materials Research Society, pp. 149–154, 1997.

[81] Wu, S., Zhu, D., Zhang, X. & Huang, J., 'Preparation and melting/freezing characteristics of Cu/Paraffin nanofluids as phase-change material (PCM)', *Energy and Fuels*, vol. 24, pp. 1894–1898, 2010.

[82] Wu, S., Zhu, D., Li, X., Li, H. & Lei, J., 'Thermal energy storage behavior of $Al_2O_3\_H_2O$ nanofluids', *Thermochimica Acta*, vol. 483, pp. 73–77, 2009.

[83] Xiao, M., Feng, B. & Gong, K., 'Thermal performance of a high conductive shape-stabilized thermal storage material', *Solar Energy Materials and Solar Cells*, vol. 69, pp. 293–296, 2001.

[84] Xie, H., Wang, J., Xi, T. & Liu, Y., 'Thermal conductivity of suspensions containing nanosized SiC particles', *International Journal of Thermophysics*, vol. 23, no. 2, pp. 571–580, 2002.

[85] Xuan, Y. & Li, Q., 'Heat transfer enhancement of nanofluids', *International Journal of Heat and Fluid Flow*, vol. 21, pp. 58–64, 2000.

[86] Yu, W., Xie, H., Li, Y. & Chen, L., 'Experimental investigation on thermal conductivity and viscosity of aluminum nitride nanofluid', *Particuology*, vol. 9, pp. 187–191, 2001.

[87] Zhang, X., Gu, H. & Fujii, M., 'Effective thermal conductivity and thermal diffusivity of nanofluids containing spherical and cylindrical nanoparticles', *Journal of Applied Physics*, vol. 100, no. 4, pp. 1–5, 2006a.

[88] Zivkovic, B. & Fujii, I., 'An analysis of isothermal phase change of phase change materials within rectangular and cylindrical containers', *Solar Energy*, vol. 70, pp. 51–61, 2000.

[89] Zhou, S. Q. & Ni, R., 'Measurement of the specific heat capacity of water-based Al_2O_3 nanofluid', *Applied Physics Letters*, vol. 92, p. 093123, 2008.

[90] Zhu, H., Lin, Y. & Yin, Y., 'A novel one-step chemical method for preparation of copper nanofluids', *Journal of Colloid and Interface Science*, vol. 227, pp. 100–103, 2004.

[91] Zhu, H. T., Zhang, C. Y., Tang, Y. M. & Wang, J. X., 'Novel synthesis and thermal conductivity of CuO nanofluid', *Journal of Physical Chemistry C*, vol. 111, no. 4, pp. 1646–1650, 2007.

3 Synthesis and Characterization Techniques

CONTENTS

3.1 INTRODUCTION

Nanoparticles of different sizes (1 nm to 100 nm) are used to make nanoparticle suspensions in order to synthesize the nanoparticle-based PCMs. Because of this, the process used to synthesize NP is highly dependent on the size and shape of the fluid containing NP. Using varied sizes and shapes will lead to variable surface area, resulting in a difference in thermal conductivity between the PCMs.

Copper oxide (CuO) nanoparticles are now being used in a wide range of applications, including catalysts, magnetic storage media, batteries, and solar energy conversion. Due to a significant effort spent on CuO nanoparticle application, there has been resurgence in basic physical property knowledge, as well as improved product performance in varied applications. Sol-gel, precipitation, sonochemical processes, microwave irradiation, and mechanical milling methods have all been developed to synthesize CuO nanoparticles.

There are numerous techniques of CuO nanoparticles preparation, but only hydrothermal and precipitation are used. These two synthesis processes are not only safe and environmentally friendly, but also rather simple. With respect to these two

DOI: 10.1201/9781003163633-3

technologies, precipitation is more popular due to its simplicity, as well as its ability to be mass-produced and inexpensive. It is possible to make different-shaped, including panicle, spherical, spindly, and rod-shaped, CuO nanoparticles using only water and NaOH without any extra surfactants in the precipitation technique.

Enlarging the processing time for similar sized grain increases the proportion of Anatase and Brookite to Rutile. All three phases of TiO_2, Anatase, Rutile, and Brookite, can be synthesized using the hydrothermal method, which allows for precise control of the process conditions. There are three polymeric forms in which TiO_2 has TiO_2 (anathrohombic) and TiO_2 (orthorhombic) with TiO_2 (Rutile) being tetragonal. Preparation of TiO_2 nanoparticles using the sol-gel process yields an isotropic yet metastable structure with great chemical uniformity, making it the favored method.

Because of their excellent physical properties, metal oxide nanoparticles are ideal candidates for nanofluids and other applications. ZnO NP is an important material in the preparation of NP, because ofits various physical, chemical, electrical, and surface characteristics. Direct band gap ZnO NP having a binding energy of 60 meV allows exciton transitions at ambient temperature, resulting in increased radiative recombination efficiency and decreased emission threshold voltage.

It is generally accepted that ZnO NP is a direct alternative for GaN (gallium nitride) and ITO (indium tin oxide), which both have insulating, conducting properties. It possesses an exceptionally high electrical and electrochemical stability, as well as the capacity to adjust resistance (10^{-3} to 10^{-5} Ω). Microelectronic devices, such as ZnO nanorods, are made possible by the electrical and optical properties of ZnO powders at the nanoscale. ZnO is considered a viable material for solid state gas sensors due to its widespread use in both bulk and thin film shape[1].

3.2   SYNTHESIS OF CuO

The water utilized in the experiment was bi-distilled water (2D water). This experiment utilized only analytical reagent-grade chemicals. A round-bottom flask filled with 250 mL of deionized water was used to dissolve 3.993 g of copper acetate to make an aqueous solution. It was necessary to combine the aqueous solution with a solution of glacial acetic acid (GA 4 mL) before heating it to 100°C for two hours with magnetic stirring at 700 rpm for two hours. A pH of between 6 and 7 was obtained when the solution was heated so that the desired amount of NaOH pellets was added and thereby precipitation with black color was observed quickly. Thus it confirmed the formation of CuO particles.

Next, the solution was centrifuged to remove contaminants and then was rinsed with 2D water five times. In the presence of an agate mortar, heated to 100°C for 36 hours, and pulverized, the metal was then processed to yield CuO particles. The following chemical reaction serves as the catalyst for the reaction: precipitation of CuO particles. CuO particles were made from copper acetate, which was a major precursor. The concentration of sodium hydroxide (NaOH) was a key component in determining the particle shape during the chemical process[2, 3].

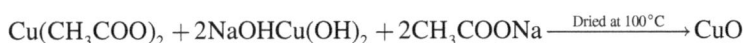

$$Cu(CH_3COO)_2 + 2NaOHCu(OH)_2 + 2CH_3COONa \xrightarrow{\text{Dried at 100°C}} CuO$$

3.3   SYNTHESIS OF TiO$_2$

TiO$_2$ NPs were synthesized by mixing titanium butoxide (titanium (II) oxide) and ethanol in a ratio of 1:4 for a sol-gel technique. A combined solution was stirred for 10 minutes at 700 revolutions per minute with a magnetic stir bar. At this point, NaOH pellets were added to the solution in a slow, steady stream, and the mixture was stirred constantly. The significant amount of white precipitation that appeared in the skies was reported after some time had passed. After that, the NPs were centrifuged to remove them from the liquid, and the purified NPs were then treated three times with 2D water. The powder was prepared by baking the topic for 24 hours at 100°C, and then grinding the sample in an agate mortar. In this synthesis, the primary precursors and reductants were titanium butoxide and sodium hydroxide (NaOH). Additionally, the inclusion of NaOH during the procedure has the potential for altering the characteristics of the NPs.

3.4   SYNTHESIS OF ZnO

ZnO NPs are prepared by dissolving 0.1M of zinc acetate in ethanol (as solvent). A magnetic stirrer made it possible to stir this solution at roughly 450 revolutions per minute for about 40 minutes. The pH was adjusted to 7 by gradually adding 5mL of sodium hydroxide solution to the mixture in drops. After that, the solution turned milky, so ethanol was added and the mixture was stirred three times before being used. It was then baked in an oven at 100°C for 32 hours to allow it to dry completely. The powder sample was finally pulverized using agate mortar, after which it was blended[4].

3.5   PARTICLE SIZE ANALYZER

In order to increase the efficiency of materials processing, characterization of NP and NPs embedded in PCMs is required. Identifying the size, shape, and crystalline structure of the synthesized NPs assists in determining their formation. Aside from that, these materials' chemical stability, phase change temperatures, latent heats, thermal conductivity, and viscosity are all given as proposed in the characterization of nanomaterials.

An instrumental for finding out the location of various sizes of particles in a liquid sample is called a particle size analyzer (PSA). Malvern Zeta Sizer determined the distribution of the various particle sizes (Malvern Instrument Inc., London, UK). These devices, a He-Ne laser (with a power output of 4 mW, wavelength 633 nm) and an avalanche photodiode detector, are both utilized in the PSA. The detector position is at 173°, and the PSA measures the scattering information at an angle close to 180°. Additionally, there are various advantages to performing this: (i) Incident beam measurements don't require samples to be transited because the backscatter is measured. Higher concentrations of the sample can be determined as light travels along a shorter path length in the sample. (ii) In comparison to sample size, contaminants such as dust particles found in dispersants tend to be larger, with scatter being minimal at 180°. When it comes to scattering, larger particles move toward the source of the energy. Therefore, dust has a negative influence when it is measured with backscatter[5].

3.6　TEM

The transmission electron microscopy (TEM) is a microscope that uses electrons to offer information on the morphology, composition, and crystallography of nanoscale objects. Conventional TEMs use thin samples irradiated with highly powerful electron beams. The intensity of the electrons that are transported through the sample is redistributed into a picture that displays the sample's morphology. An image is formed using an intense and concentrated electron beam rather than visible light on a fluorescent screen to illuminate a sample that is nano in size. The arrangement of atoms in the specimen, and their degree of order, are visible from TEM images.

A laser beam focused on the sample can turn the image into a real thing. TEM resolutions can be described as high-resolution. The result of this experiment displays the fact that the material is crystalline. Crystallinity can also be checked using electron diffraction of the material. In order to do an energy dispersive X-ray (EDX) examination on the TEM, an EDX analyzer was added to the setup. The EDX method distinguishes the elemental makeup of any materials that have an atomic number greater than boron. The X-ray fluorescence (XRF) imaging procedure requires that the X-ray scanning beam strikes the atoms that are under study. It is characteristic of the element that generated it, now as X-rays[5].

To verify the size of the synthesized NPs which were placed between 1 nm and 100 nm, TEM was utilized. The CM120 series TEM of Philips is ideal for determining the size and surface area of NPs. A high-voltage transmission electron microscope, CM120, can reach voltages up to 120kV. Samples of colloidal dispersion were made on a copper grid coated in a carbon sheet. For removal, the solvent was evaporated after the dispersion process was complete.

3.7　SEM

The morphology of the developed NPs was examined using field emission scanning electron microscopy (FESEM, LEO 1530, Zeiss, Germany). Instead of light, the scanning electron microscope employs a stream of electrons that is well-aligned to make an image. Heating a metal filament on the top of a microscope creates an electron beam known as the electron gun. A vertical path is traced by the microscope's column as an electronic beam follows it.

The secondary electrons are ejected as the beam impacts the sample. When secondary electrons are gathered, they are then converted into a signal that can be detected. On a computer screen, the image that is so created is observed. As a result, the electron gun has a field emission cathode that provides an extremely narrow probing beam with both strong and fragile electron energy. By adjusting the magnification, a higher resolution of the space is obtained. X-ray (EDX) setup was occasionally used to analyze the elemental composition of NPs while doing a scanning electron microscopy (SEM) investigation.

SEM employs a scanning electron beam to capture electrons that have been scattered by the material. The backscattered electron image does not need forward-transmitted signals, which reduces the amount of electron beam energy required. The image does not need electronic transparency, as the image is produced using backscattered signals, which require a conductivity level to prevent charging[6].

3.8 XRD

The sample's qualities, such as atomic arrangement, crystal structure, crystalline perfection, particle size, and texturing can be discovered using X-ray diffraction. An X-ray scanner is required to see within the crystals since the arrays are constructed of crystals spaced using atomic units (Å). Instead of viewing the samples while moving the machine around, the X-rays were directed to the sample from a fixed location.

When scatter light that has been reflected, there are two different possibilities. It's possible to adjust how far the light spreads by altering the size of the space between individual metal atoms, or to disperse it using a mathematical relationship (i.e. $2D \sin\theta = n\lambda$). To estimate the average crystallite size (D), the Debye-Scherrer relation was used[7].

$$D = 0.9 \ \lambda/W \ \cos\theta$$

where, D—mean particle size
λ—wavelength of X-ray
W—full width half maximum
θ—scattering angle.

X-ray diffraction was used to examine the three samples and all results were obtained using a Rigaku Miniflex-II C model diffractometer. This was an area of coverage that ranged from 20 to 80°.

3.9 FTIR

It is a non-destructive method for materials analysis, also known as infrared spectroscopy. The study on infrared radiation interacts with materials depending on the frequency of the photons and is known as infrared absorption spectroscopy. Both organic and inorganic materials can be studied using FTIR to learn about chemical bonding and molecular structure. Infrared (IR) has three main regions: near-IR ($400–10 \ cm^{-1}$), mid-IR ($4000–400 \ cm^{-1}$), and far-IR ($14000–4000 \ cm^{-1}$). The photons in the infrared spectrum are capable of shaking up groupings of atoms, which are known as optical biophotonics. Likewise, the infrared absorption bands for molecules and electronic transitions are discrete energy and frequency states.

The vibrational modes of chemical bonds change in a predictable manner, and they can absorb infrared radiation at frequencies that correspond to the original vibrational modes. When it comes to the spectrum created by the function of frequency on the absorption of radiation, functional groups and compounds can be found. Different forms of impurities have different infrared spectral signatures. Impurity concentration and host material bonding are analyzed and calculated using spectral measurements on these spectral bands, which are provided by these spectral measurements. It is essential that the measured interferogram signal is translated by a trained interpreter before the signal can be used to identify an object. Using individual frequencies requires a method of "decoding." The Fourier transformation can be used to approximate this calculation. To execute this transformation, the computer uses the algorithm to transform the data, and then displays the information for analysis. IR absorption experiments in the spectrum range of 4000–400 cm^{-1} were conducted using a Perkin Elmer FTIR spectrometer. This instrument's spectral resolution is 4 cm^{-1}. IR investigations were done using KBr powder and ferrite nanoparticle-encapsulated pellets.

3.10   LASER FLASH ANALYZER

Xenon flash lamps replace the laser in this well-proven method and lasers are less commonly utilized. The inbuilt sample changer for four samples offers the ability to automatically perform measurements on many samples. When performing a measurement, use an insulated holding plate and a short-energy pulse to heat the sample in a plane parallel to the plate. Finally, an infrared detector is used to measure the top surface temperature change. Thermal diffusivity has a direct influence on the signal, rising the higher the thermal diffusivity[8].

$$k(T)-\alpha(T).C_p(T).\rho(T)$$

where, α—thermal diffusivity
ρ—density
C_p—specific heat capacity
k—thermal conductivity
T—temperature

The sample's calculated thermal conductivity is determined from the formula previously mentioned. Thermal conductivity measurements using guarded hot plates, heat flow meters, or thermal conductivity testers can take significantly less time than LFA studies. It is possible to determine the thermal conductivity of the samples from room temperature to 300°C.

3.11   VISCOMETER

As the two layers of fluid are forced to flow against one another, they create internal friction, which is measured in viscosity. Measurements of viscosity were made using a Brookfield Ultra Programmable Viscometer. The DV-III Ultra's method of operation is to move a spindle through a calibrated spring submerged in the test fluid. Measurement of the viscous drag is done by using a spring to gauge the amount of deflection it causes. The rotational transducer measures spring deflection throughout the spring's motion. This parameter is set in units of cP (centipoises) or mPa-s (milliPascal-second), with the spindle's rotational speed, the size and form of the spindle, and the calibrated spring influencing the setting.

The fluid must be contained in a container in order to be weighed or measured. Autozeroing of a viscometer must be performed before readings are taken from each power supply. The DV-III Ultra standard spindles are built to work with a Griffin beaker. It can measure fluids over a large range, such as those on the RVDV-III+ which range from 100 cP to 40,000,000 cP. Deflection of the calibrated spring is used to measure the viscous drag of the fluid on the spindle.

The data logger records data on the shear rate, shear strain, and viscosity at room temperature. This procedure generally involves trial and error to pick a spindle and rotation speed for an unknown fluid. In the current investigation, the SC-21 type spindle was adjusted to 250 revolutions per minute (RPM) for the tests. To accurately control the temperature of the sample, a thermostatic bath should be used. To ensure accuracy, the viscometer was calibrated using the standards. Test data repeatability was determined to be within ±0.2% of the original test data[9].

3.11.1 THERMOGRAVIMETRIC ANALYSIS

Analysis using thermographic techniques measures a material's thermal stability as well as the amount of volatile components released when it is heated. A standard procedure includes using helium or argon in the gas mixture and recording weight as a function of temperature.

Thermogravimetric analysis (TGA) has the capacity to offer information such as the components of a multi-component system, the thermal stability of materials, the oxidative stability of materials, an estimation of the lifetime of a product, the breakdown kinetics of materials, moisture and volatile content of materials. The basic elements that are required to make this invention are a crucible that is used to contain the sample, a furnace that is capable of heating the sample to a high temperature, and a balance that continuously monitors the weight of the sample. The analytical balance is located outside the furnace chamber, and the test sample is placed in an alumina cup. Nanomaterial embedded PCMs were also tested, with a heating rate of 20°C/min and a cooling rate of 1 bar using nitrogen as the coolant. A total of 6.32 mg of samples were stored in an airtight aluminum pan.

3.12 DSC

A powerful tool for measuring the heat capacity of nanomaterials (nanofluid PCMs) is Differential Scanning Calorimetry (DSC). To determine the quantity of heat flow required to raise the reference pan to a specific temperature, the two pans are heated to the same temperature and the quantity of heat flow is calculated as a function of temperature. The observance samples and reference containers at roughly the same temperature is possible since the experiment pans are nearly identical in temperature. The differential in heat flow is measured to determine the sample's heat capacity. In this case, the pan has experienced a phase transition and consequently the temperature has risen or fallen by a greater amount than the reference pan. This results in an endothermal or exothermal peak on the DSC curve, which refers to melting or freezing, respectively.

The results of individual measurements will fluctuate substantially if the heating rates and sample masses are varied. The measurement configuration results are very consistent across one set of measurements; however, the results are not the same when different measurement configurations are used. The heating rate and sample mass do not appear to affect the beginning and end of the peaks, respectively.

At a rate of 5°C/min, the base PCMs and nanomaterials-incorporated PCMs were subjected to heat flow tests. Samples were sealed in an aluminum pan that contained 6.5 mg of mass. This finding demonstrates that the melting and solidification points are fixed and do not change, even when temperature increases. The integrated heat was obtained by using software on the DSC equipment to carry out the integration[10].

3.13 THERMAL CYCLE TESTER

Thermal cyclers are used to test the durability of the samples. Cutting-edge thermoelectric refrigeration system was employed. Temperature changes occur at a rate of 4°C per second. Temperature can be increased or decreased to suit the needs of the experiment. This test was performed on the PCMs to find out if they can last for many heat cycles by running them through the thermal cycler (BIOER TC-25/H model).

REFERENCES

[1] Coronado, D. R., Gattorno, G. R., Pesqueria, M. E., Cab, C., Coss, R. D. & Oskam, G., 'Phase-pure TiO2 nanoparticles: Anatase, brookite and rutile', *Nanotechnology*, vol. 19, pp. 45605–45614, 2008.

[2] Gunther, E., Hiebler, S., Mehling, H. & Redlich, R., 'Enthalpy of phase change materials as a function of temperature: Required accuracy and suitable measurement methods', *International Journal of Thermophysics*, vol. 30, pp. 1257–1269, 2009.

[3] Jeng, J.-Y., Liu, J.-C. & Jean, J.-H., 'Dispersion of oleate-modified CuO nanoparticles in a nonpolar solvent', *Journal of the American Ceramic Society*, vol. 90, pp. 3676–3679, 2007.

[4] Li, Y., Yu, J., Liu, Y., Huang, R., Wang, Z. & Zhao, Y., 'A review on removal of mercury from flue gas utilizing existing air pollutant control devices (APCDs)', *Journal of Hazardous Materials*, p. 128132, 2022.

[5] Jingfa, D., Qi, S., Yulong, Z., Songying, C. & Dong, W., 'A novel process for preparation of a Cu/ZnO/Al₂O₃ ultrafine catalyst for methanol synthesis from $CO_2 + H_2$: Comparison of various preparation methods', *Applied Catalysis A: General*, vol. 139, pp. 75–85, 1996.

[6] Meng, X., Lee, J. H., Park, M. H., Yu, S. M., Shin, D. W., Yang, C., Bhoraskar, V. N. & Yoo, J. B., 'Anatase-assisted growth of nanoscale crystalline brookite film and rutile-rods on SiO₂/Si substrate in a single hydrothermal process', *Cryst Eng Comm*, vol. 13, pp. 3983–3987, 2011.

[7] Meulenkamp, E. A., 'Synthesis and growth of ZnO nanoparticles', *Journal of Physical Chemistry B*, vol. 102, no. 29, pp. 5566–5572, 1998.

[8] Monticone, S., Tufeu, R. & Kanaev, A. V., 'Complex nature of the UV and visible fluo-rescence of colloidal ZnO nanoparticles', *Journal of Physical Chemistry B*, vol. 102, no. 16, pp. 2854–2862, 1998.

[9] Wu, S., Zhu, D., Zhang, X. & Huang, J., 'Preparation and melting/freezing character-istics of Cu/Paraffin nanofluids as phase-change material (PCM)', *Energy and Fuels*, vol. 24, pp. 1894–1898, 2010.

[10] Zaban, A., Aruna, S. T., Tirosh, S., Gregg, B. A. & Mastai, Y., 'The effect of the prepara-tion condition of TiO_2 colloids on their surface structures', *Journal of Physical Chemistry B*, vol. 104, no. 17, pp. 4130–4133, 2000.

4 Phase Change Analysis

CONTENTS

4.1 INTRODUCTION

Mathematical modeling was used to simulate the PCMs embedded with the nano-structures' solidification and melting phases in this chapter. The state (solid/liquid) of the PCMs is subject to change with respect to their phase change temperature. When solidification and melting occur inside a spherical enclosure in one dimension, the quasi-steady model is used to investigate the impact of NPs on the crystallization and melting processes. This model was constructed and compared to pure PCMs in a spherical encapsulation to assess whether there were differences in results. During solidification and melting of spheres, analytical methods are presented for detecting the interface position at specific time steps. Solutions show that the total solidification time depends on the Stefan number and the heat generation parameter, but they also show that this time can change.

Polyethylene capsules can be filled with PCMs, which are then encapsulated inside the capsules. Encapsulation in spherical containers is most often used because of the positive relationship between the amount of energy stored per unit volume and the surface available for heat transfer. The main attribute for controlling the migration of two phases in a reaction vessel is knowing the position of the surface where the phase change occurs. This surface releases or absorbs heat, making it difficult to solve the problem.

These assumptions predicted the transient location of the PCM in the spherical capsule[1]:

1. A surface formed instantly by the creation of a solid-liquid contact is said to be instantaneously formed and to be moving through the liquid phase.
2. Both the volume and density of the pure PCMs and PCMs embedded with nanostructures remain constant when used with both frozen and aqueous phases, respectively.
3. A properly insulated PCM tank prevents heat transfer fluid from expanding, making the flow incompressible in the tank.
4. As temperatures rise, the PCMs and nanostructures-incorporated PCMs have a different range of temperatures.

DOI: 10.1201/9781003163633-4

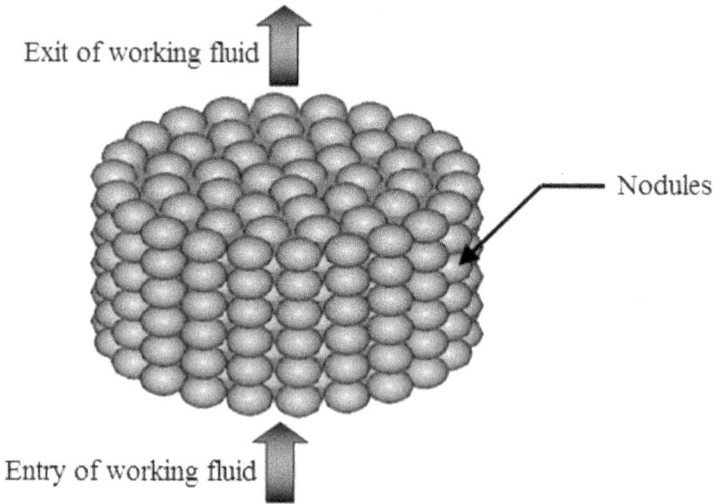

FIGURE 4.1   Arrangement of spherical capsules in LTES systems[2].

5. These temperature-dependent thermophysical features of the heat transfer fluid have minor impact on the overall heat transfer fluid system
   performance.
6. It is practically negligible whether heat is transferred between the PCMs
   and the PCMs containing the nanostructures capsules via radiation.
7. At constant temperature, the PCMs that are free of nanostructures as well
   as those embedded within the capsule endure a phase change.

Over the last few years, various studies pertaining to PCMs' thermal performance
and phase change behavior have been conducted in packed bed LTES systems using
either spherical or cylindrical-packed bed plates. Due to their wide range of applications and increased material capacity, the spherical capsules known as nodules in
Figure 4.1 were selected.

4.2   SOLIDIFICATION ANALYSIS

The typical behavior of the spinning boundary is studied using a single spherical
capsule with a defined radius, R. The PCM's temperature, Ti, is initially above the
PCM's fusion temperature (T_f). According to Figure 4.2, the surface of the sphere
that is at radius $r = R$ is kept at a constant temperature of $To < Tf$. With a constant
volumetric rate, energy is created around the entire sphere (g).

As the sphere's surface begins to solidify, a front-moving solidification occurs in
the liquid and then drives itself into the liquid. This is a symmetric problem that can
be solved from 0 to R on the moving surface by setting the variable to the coordinate
of the moving surface.

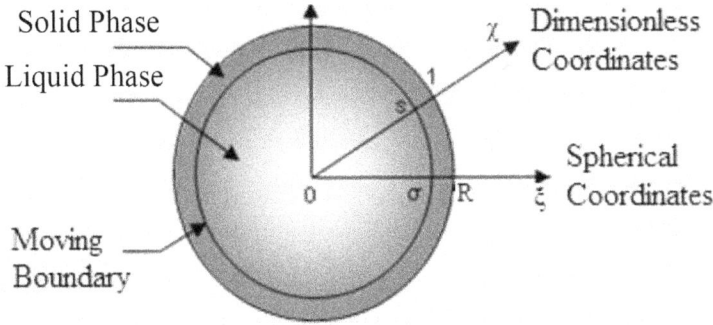

FIGURE 4.2   Model of the spherically encapsulated PCM.

At fusion, the sphere will be at temperature Ti when it has reached its highest possible temperature. Heat generation within the sphere is accounted for when analyzing the whole effect of the sphere on its surroundings. The symmetry of the equation allows for an analysis of half of the sphere.

The governing heat equations for solidification are based on the previous assumptions[3].

$$\alpha\left(\frac{\partial^2 T_s}{\partial r^2} + \frac{2 \partial T_s}{r \partial r}\right) + \frac{g}{\rho c_p} = \frac{\partial T_s}{\partial t}, \sigma \leq r \leq R \tag{4.1}$$

$$\alpha\left(\frac{\partial^2 T_L}{\partial r^2} + \frac{2 \partial T_L}{r \partial r}\right) + \frac{g}{\rho c_p} = \frac{\partial T_L}{\partial t}, 0 \leq r \leq \sigma \tag{4.2}$$

The product of thermal diffusivity (D), density (ρ), specific heat (C_p), and subscripts of S and L indicate the presence of solid and liquid phases in the system.

The boundary conditions are

1. $T_S(r = R, t) = T_O$
2. $T_S(r = \sigma, t) = T_f$
3. $T_S(r = \sigma, t) = T_f$
4. $\dfrac{\partial T_L(r = 0, t)}{\partial r} = 0$

The initial conditions are as follows:

1. $T_L(r, t = 0) = T_i$
2. $\sigma(t = 0) = R$

In an interface between a solid and a liquid, the interface energy balance is determined by

$$k\frac{\partial T_S(r=\sigma,t)}{\partial r} - k\frac{\partial T_L(r=\sigma,t)}{\partial r} = \rho H\frac{d\sigma}{dt} \tag{4.3}$$

where "k" is the thermal conductivity, "H" is the latent heat of fusion, and "s" is the solidification front position.

Dimensionless variables are used, and the governing equation, boundary conditions, starting conditions, and the energy balance equation at the interface are all made dimensionless as a result of their usage.

The dimensionless quantities are defined as:

$$x = \frac{r}{R}, \theta_s = \frac{T_f - T_S}{T_f - T_o}, \theta_L = \frac{T_f - T_L}{T_f - T_o}$$

$$s = \frac{\sigma}{R}, F_o = \frac{\alpha t}{R^2}, \beta = \frac{R^2 g}{k(T_f - T_o)} \tag{4.4}$$

Figure 4.2 depicts a ball with the two coordinate systems, the dimensional and the dimensionless. When non-dimensional quantities (4.4) are substituted in governing equations (4.1) and (4.2), they become the non-dimensional governing equations and they are as follows

$$\frac{\partial^2 \theta_S}{\partial x^2} + \frac{2\partial\,\theta_S}{x\partial x} - \beta = \frac{\partial\,\theta_S}{\partial F_o}, s \le x \le 1 \tag{4.5}$$

$$\frac{\partial^2 \theta_L}{\partial x^2} + \frac{2\partial\,\theta_L}{x\partial x} - \beta = \frac{\partial\,\theta_L}{\partial F_o}, 0 \le x \le s \tag{4.6}$$

The non-dimensional boundary and initial conditions are similarly specified:

$$\theta_S(x=1,F_o)=1,$$

$$\theta_S(x=s,F_o)=0,$$

$$\theta_L(x=s,F_o)=0,$$

$$\frac{\partial\theta_L(x=0,F_o)}{\partial x}=0$$

$$\theta_L(x,F_o=0)=\frac{T_f-T_i}{T_f-T_o}$$

$$s(F_o=0)=1$$

The interface energy equation (4.3) transforms into

$$Ste\left[\frac{\partial\theta_s(s,F_o)}{\partial x} - \frac{\partial\theta_L(s,F_o)}{\partial x}\right] = -\frac{ds}{dF_o} \tag{4.7}$$

where, "*Ste*" is the Stefan number and it is defined as $Ste = \frac{(T_f - T_o)c_p}{k}$.

It's assumed that energy generation parameter β governs the development of a solidification solution for the differential equation discussed earlier. Assumption of quasi-stationary is appropriate for this type of issue. The quasi-steady model is accurate for low values of the Stefan number. The interface energy equation is preserved in the model.

The distribution of temperature and the mobility of the contact both change over time. As the interface moves in a steady state, temperature is dispersed across it in an instant. A solution obtained by neglecting the variables that are only defined for a moment, integrating equations (4.5) and (4.6), and applying boundary conditions.

$$\theta_S(x,F_o) = \frac{6s + \beta s(s-1)}{6(s-1)} + \frac{6 + \beta(s^2-1)x}{6(1-s)} + \frac{\beta x^2}{6}, s \le x \le 1 \tag{4.8}$$

$$\theta_L(x,F_o) = \frac{\beta(x^2-s^2)}{6}, 0 \le x \le s \tag{4.9}$$

Differentiating equations (4.8) and (4.9) and substituting the differentials into the interface equation (4.7), and applying the initial condition $s(F_o = 0) = 1$, results in

$$Ste\int_0^{F_o} dF_o = \frac{6}{\beta}\int_1^s \frac{(s-1)ds}{s^2 + \left(\frac{6}{\beta}-1\right)} \tag{4.10}$$

For three circumstances, the integral must be assessed.
    Case 1: $\beta < 6.0$

$$SteF_o = \left(\frac{6s}{\beta} - \frac{6}{\beta}\right) + \left(\frac{6}{\beta}\sqrt{\frac{6}{\beta}} - 1\tan^{-1}\left(\frac{1}{\sqrt{\frac{6}{\beta}-1}}\right)\right) - \left(\frac{6}{\beta}\sqrt{\frac{6}{\beta}} - 1\tan^{-1}\left(\frac{1}{\sqrt{\frac{6}{\beta}-1}}\right)\right)$$

$$+ \frac{5.375}{\beta} - \frac{3}{\beta}\ln(6 - \beta + \beta s^2)$$

For $\beta < 6$, the entire sphere is solidified at Fo_{sol}, before the steady state is reached. Since sensible heat is released by the PCM and its heat generation occurs, the PCM's temperature continues to reduce even after the entire solidification process is complete. Hence the previous solution is applicable when $Fo < Fo_{sol}$. Complete solidification time of spheres can be calculated by,

$$SteFo_{sol} = \left(-\frac{6}{\beta}\right) + \left|\frac{6}{\beta}\sqrt{\frac{6}{\beta}-1}\tan^{-1}\left(\frac{1}{\sqrt{\frac{6}{\beta}-1}}\right)\right| + \frac{5.375}{\beta} - \frac{3}{\beta}\ln(6-\beta)$$

Case 2: $\beta > 6.0$

$$Ste\int dF_o = \frac{6}{\beta}\int_0^1 \frac{s^2}{s^2-\left(\frac{\beta-6}{\beta}\right)}ds - \int_0^1 \frac{6s}{6-\beta+\beta s^2}ds$$

$$SteFo = \frac{6}{\beta}(s-1) + \frac{3}{\beta}\sqrt{\frac{\beta-6}{6}}\ln\left|\frac{\sqrt{\beta}s-\sqrt{\beta-6}}{\sqrt{\beta}s+\sqrt{\beta-6}}\right| - \frac{3}{\beta}\ln\left(1-\frac{\beta}{6}+\frac{\beta}{6}s^2\right)$$

Even when the equilibrium criterion is met, the sphere will not completely harden. Instead, it maintains constant steady-state conditions and goes to a precise interface location, known as S_{ST}, where all the energy created is transmitted into the water in the temperature being kept constant.

$$s_{ST} = \sqrt{1-\frac{6}{\beta}}$$

Case 3: $\beta = 6.0$

$$SteF_o = \int_0^1 \frac{6s(s-1)}{6s^2}ds$$

$$SteF_o = (s-1) - \ln s$$

In this particular instance, when the steady state is reached, the sphere will entirely solidify. When the interface is in equilibrium, the energy is transmitted to the cool surface.

4.3 MELTING ANALYSIS

When the PCM is at the phase change temperature, the PCM within the spherical encapsulation is initially solid (T_f). The temperature on the surface of the sphere is maintained at $T_o > T_f$. The melting of PCM begins on the surface and proceeds toward the interface. The partial melting of the solid phase causes the creation of a solid-liquid zone instead of a solid phase. Using a quasi-steady approximation method, it is assumed that the mixture evolves and the solid phase's temperature rises instantly to the fusion temperature. The fusion temperature will be reached when the mixture has been thoroughly incorporated. But the liquid content in the mixture grows as a result of the additional energy and sensible heat supply provided by the heat surface. When the entire sphere melts, the temperature of the liquid will increase until it reaches the stable temperature for the system.

The Rayleigh number has a large effect on the amount of natural convection. This problem can be solved by defining heat transfer as the principal mode and include convective effects in the liquid's thermal conductivity. Correlation is used to determine PCM's effective thermal conductivity and which is given here,

$$\frac{k_{eff}}{k_L} = 0.202 \, Ra_L^{0.228} \left(\frac{L}{r_i}\right)^{0.252} Pr^{0.029} \tag{4.11}$$

Where k_{eff}—effective thermal conductivity, W/mK
k_L—thermal conductivity of the liquid PCM, W/mK

In the dimensionless formulation, the energy generation parameter β, Stefan number Ste, and dimensionless temperature θ_L are redefined as,

$$\beta = \frac{R^2 g}{k_{eff}(T_0 - T_f)}, \; Ste = \frac{c(T_0 - T_{L-s})}{H}, \; \theta_L = \frac{T_0 - T_L(x,t)}{T_0 - T_f} \tag{4.12}$$

For the liquid phase, the transient heat equation with energy generation is

$$\alpha\left(\frac{\partial^2 T_L}{\partial r^2} + \frac{2\partial \, T_L}{r\partial r}\right) + \frac{g}{\rho c_p} = \frac{\partial T_L}{\partial t} \tag{4.13}$$

The boundary conditions and initial condition include,

1. $T_L(r = R,t) = T_O$
2. $T_L(r = \sigma,t) = T_f$
3. $\sigma(t = 0) = R$

The dimensionless heat equation is obtained from equation (4.13), and the corresponding boundary and initial conditions are

$$\frac{\partial^2 \theta_L}{\partial x^2} + \frac{2\partial\, \theta_L}{x\partial x} - \beta = \frac{\partial\, \theta_L}{\partial F_o}, s \leq x \leq 1 \tag{4.14}$$

1. $\theta_L(x=1, F_o) = 0$
2. $\theta_L(x=s, F_o) = 1$
3. $s(F_o = 0) = 1$

The transient terms from the governing equation (4.13) can be ignored if quasi-stationarity is assumed, and the differential equation can be solved.

$$\theta_L = \frac{-\beta - 6s + \beta s^3}{6(1-s)} + \frac{6s + \beta s - \beta s^3}{6(1-s)x} + \frac{\beta x^2}{6} \tag{4.15}$$

When solids are melted, a combination of solid and liquid phases is formed instead of a pure solid phase due to partial melting. Thus, the interface energy equation is newly redefined, having a term called γ added to it.
    The mixture's energy conservation is represented by

$$g = \rho H \frac{d\gamma}{dt}$$

Integrating this equation and applying the initial condition $\gamma(0) = 0$, it transforms into

$$\gamma = \frac{g}{\rho H} t$$

Assuming that no heat is transported through the mixture through conduction, the interface energy equation becomes

$$-k\frac{\partial T_L}{\partial r}(r = \sigma, t) = \frac{\partial \sigma}{\partial t} \rho H (1 - \gamma)$$

We can acquire the dimensionless interface energy equation by solving the governing equation and substituting it in it.

$$SteF_O = \frac{1 - 3s^2 + 2s^3}{6 + \beta - 3\beta s^2 + 2\beta s^3} \tag{4.16}$$

The sphere's total melting time can be calculated using (i.e. $s = 0$, $F_o = Fo_{mel}$),

$$SteF_{O_{mel}} = \frac{1}{6 + \beta} \tag{4.17}$$

Nanomaterials-infused PCMs have higher thermal conductivity and single-phase heat transfer coefficients than base fluids, despite the fact that base fluids have higher thermal conductivity and single-phase heat transfer coefficients. Increased heat transfer coefficient is only a contributing factor to the overall increased thermal conductivity. NP mobility is usually liable for increasing turbulence and heat dispersion in this phenomenon. This includes inertia, Brownian diffusion, thermophoresis, diffusiophoresis, Magnus effect, fluid drainage, and gravity. Nanofluid characteristics can be correctly predicted using a variety of models, but the most widely accepted model is the Hamilton-Crosser model. The thermal conductivity of the surface has thus shifted, and as a result, the heat generation parameter β fluctuates. Despite the dispersion and homogeneity of NPs, the properties of PCMs with NP dispersion are not significantly different from those of pure PCMs. However, the time required for entire solidification and the melting time for a specific configuration are greatly affected[4].

These assumptions were used to predict the transient interface position in the spherical capsule.

1. The phase change occurs instantaneously and flows though the liquid phase.
2. Changes in the volume of PCM and the density of nanofluids in both solid and liquid phases stay the same during freezing.
3. There is good insulation in the tank, and the flow of heat transfer fluid is incompressible.
4. The nanofluids and heat transfer fluid temperatures vary in the tank's axial direction.
5. Thermophysical features of the heat transfer fluid in heat transfer equipment has a negligible effect on its temperature.
6. The heat transfer between the capsules is low due to radiation.
7. As the nanofluids in the capsule maintain a steady temperature, they experience a phase change.

The approximate thermal conductivity of the implanted PCMs is derived from the Hamilton-Crosser model by using nanomaterials of the appropriate dimensions[5],

$$\frac{k_{nf}}{k_f} = \frac{k_s + (n-1)k_f - (n-1)(k_f - k_s)\varphi}{k_s + (n-1)k_f + (k_f - k_s)\varphi} \tag{4.18}$$

The density of the nanomaterials embedded PCMs is given by

$$\rho_{nf} = (1-\varphi)\rho_f + \varphi\rho_s \tag{4.19}$$

The new dimensionless heat generation parameter β_{nf} is given by the relation,

$$\beta_{nf} = \frac{R^2 g}{k_{nf}(T_0 - T_f)} \tag{4.20}$$

Using ρ_{nf} in the governing equation and solving the partial differential equation,

$$Ste \int_0^{F_O} dF_{O_{nf}} = \frac{6}{\beta_{nf}} \int_1^s \frac{(s-1)ds}{s^2 + \left(\frac{6}{\beta_{nf}} - 1\right)} \tag{4.21}$$

F_{onf} would replace F_o, and the results obtained are nearly identical to earlier results, except Fonf would take the place of F_o and β_{nf} would substitute for β. Similarly, in place of ρ and k, melting ρ_{nf} and k_{nf} are used. Similarly, the heat generation parameter and the entire melting time change. Melt time is determined by the relation,

$$SteF_{O_{nf}} = \frac{1}{6 + \beta_{nf}} \tag{4.22}$$

Through the use of MATLAB, the process was accelerated and a complete closed network between the models was built. Models connected through MATLAB are each programmed to be incorporated into simulation analysis as they are passed through the system. The thermophysical characteristics model was utilized in combining the freezing and melting models. This is designed to more clearly show the effects of the mass fraction of NPs on the charging and discharging cycles for PCMs with integrated nanostructures.

In addition, the simulation was utilized to explore the formation and propagation of the solid-liquid interface layer of PCMs containing nanostructures, as well as the freezing and melting of the nanostructures. Time-dependent governing equations and mathematical relations had to be coded and networked in order to conduct simulation analysis.

REFERENCES

[1] Chan, S. H. & Hsu, K. Y., 'Application of a generalized phase change model for melting and solidification of material with internal heat generation', In: *Proceedings of the AIAA 19th Thermophysics Conference*, Snowmass, CO, pp. 25–28, 1984.
[2] Kalaiselvam, S., Parameshwaran, R. & Harikrishnan, S., 'Analytical and experimental investigations of nanoparticles embedded phase change materials for cooling application in modern buildings', *Renewable Energy*, vol. 39, pp. 375–387, 2012.
[3] Pedroso, R. I. & Domoto, G. A., 'Domoto. Inward spherical solidification: Solution by the method of strained coordinates', *International Journal of Heat and Mass Transfer*, vol. 16, no. 5, pp. 1037–1043, 1973.

[4] Scanlan, J. A., Bishop, E. H. & Powe, R. E., 'Natural convection heat transfer between concentric spheres', *International Journal of Heat and Mass Transfer*, vol. 13, no. 12, pp. 1857–1872, 1970.

[5] Harikrishnan, S. & Kalaiselvam, S., 'Preparation and thermal characteristics of nano-materials embedded phase change materials for thermal energy storage', Shodh ganga, 2017. http://hdl.handle.net/10603/142206.

5 NEPCMs for Cooling Applications

CONTENTS

5.1 INTRODUCTION

In this chapter, two base fluids were employed: oleic acid and a water-glycerol mixture. It was employed as a support material for copper NPs distributed in base fluids. This combined mixture of water and glycerol has a melting point of around (6–8)°C and (3–4)°C. The latent heats in a mixture of oleic acid and water-glycerol at a room temperature are 140.16 kJ/kg and 306.42 kJ/kg.

The two-step procedure was used to prepare the nanomaterial embedded PCMs. Two pure PCMs and PCMs with NPs were evaluated for mass fraction, number of thermal cycles, and thermal properties. PCM thermal conductivity was measured using a laser flash analyzer and it was confirmed that NP dispersion increased the PCM thermal conductivity. The higher viscosity due to NPs was also investigated

DOI: 10.1201/9781003163633-5

to assess whether this change affects heat transfer of the new PCMs during heating and cooling[1].

5.2 OLEIC ACID-CUO NANOFLUID PCM

A nanofluid PCM containing CuO-oleic acid has been synthesized and characterized. In addition, the freezing and melting properties of the NPs-incorporated PCMs were explored to determine the increased thermal and heat transfer capabilities of the LTES system. Experimental and analytical studies of the transient location of nanomaterials incorporated in PCMs were conducted.

5.2.1 CHARACTERIZATION OF CuO NPS

Figure 5.1 shows the TEM images of samples whose particles varied in size from 1 nm to 80 nm. The large surface-to-volume ratios of these particles thereby enhance the heat conductivity of the base fluid (PCM). As a result, it will assist in shortening the time required for the PCM to store and release energy. Additionally, the TEM image provided evidence that clusters were forming. Figure 5.2 shows the CuO NPs' SEM investigation, which confirms their spherical shape.

Figure 5.3 displays the XRD pattern of precipitation procedure CuO NPs. This depicts the monoclinic structure of single-phase with the lattice parameter of a = 4.84 Å, b = 3.47 Å, c = 5.33 Å. The XRD pattern corresponded well with the JCPDS data (file No. 45–0937). The XRD pattern showed no peaks for impurities. The dispersion of the peaks showed that the particles were of nanoscale size. Figure 5.3 depicts the dispersion of the NPs in the base fluid as a function of the CuO particle size distribution. In CuO-oleic acid nanofluid PCMs, the particle size distributions exhibited only

FIGURE 5.1 TEM images of CuO NPs[3].

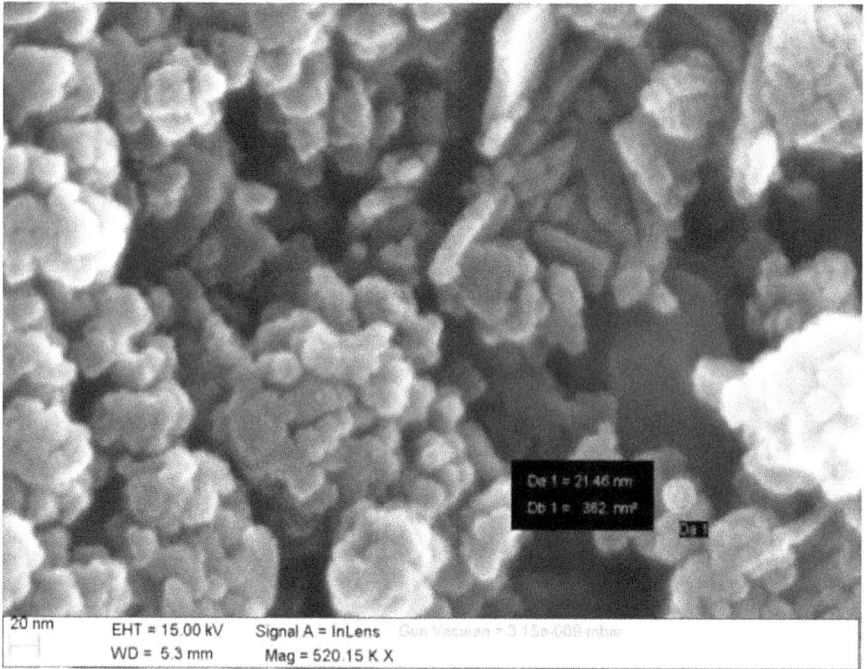

FIGURE 5.2   SEM image of CuO NPs[3].

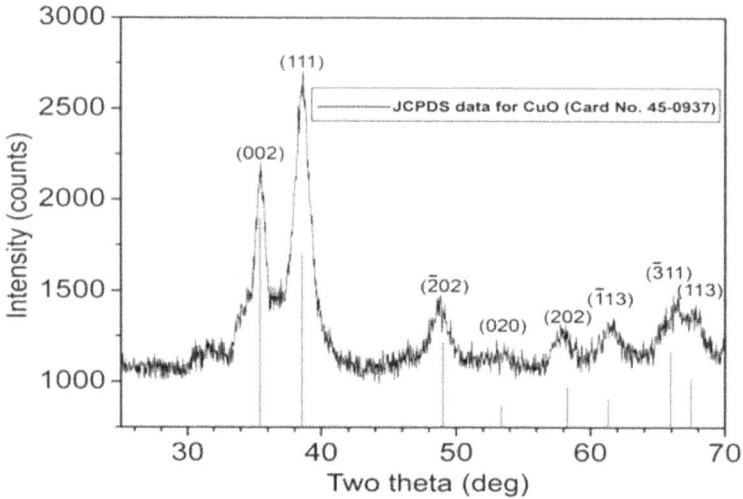

FIGURE 5.3   XRD pattern of CuO NPs[16].

Size Distribution by Volume

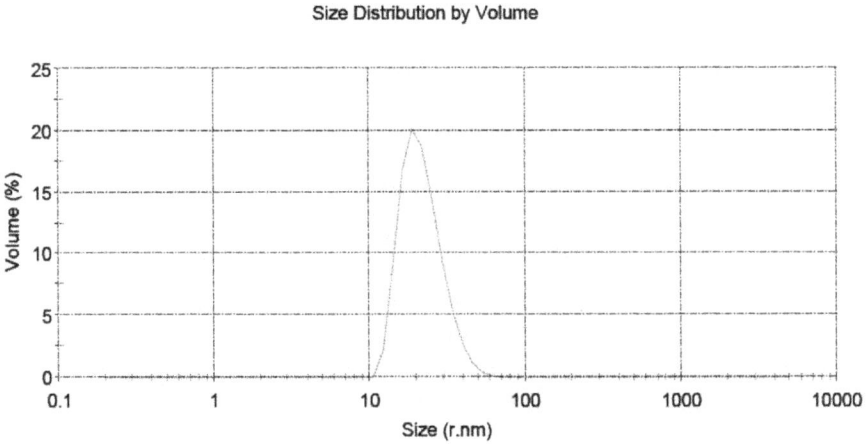

FIGURE 5.4 Particle size distributions of CuO NPs in oleic acid[16].

one peak, indicating a homogeneous distribution of CuO NPs. Figure 5.4 shows the average CuO NP particle size of 10 to 60 nm[2].

5.2.2 PREPARATION OF NANOMATERIALS EMBEDDED PCMS

Oleic acid was obtained from Thermofisher Scientific Private Limited in India. Oleic acid's thermophysical characteristics are found in Table 5.1. It is essential that nanofluid PCMs be prepared in order to improve thermal conductivity of the base fluid. The nanofluid PCMs were prepared using a two-step process. During the preparation, the fluid must be monitored carefully to ensure that the dispersion is consistent, that particle agglomeration is minimal, and that chemical changes are avoided. The PCMs' low aggregation and dispersion of nanofluids was tested using ultrasonication, as opposed to other techniques. Ultrasonic vibration was performed at a 40 kHz frequency for the various mass fractions, with the nanofluid PCM dwelling periods in the vibrator enduring 30, 35, 40, and 45 minutes for 0.5, 1.0, 1.5, and 2.0 wt% fractions, respectively[4].

TABLE 5.1
Thermophysical Properties of Oleic Acid[16]

S.No	Properties of the PCM	Values
1.	Molecular mass (g/mol)	282.46
2.	Melting temperature (°C)	6–8
3.	Density (kg/m³)	887
4.	Latent heat (kJ/kg)	140.16
5.	Specific heat capacity (kJ/kg K)	2.04
6.	Thermal conductivity (W/m K)	0.224

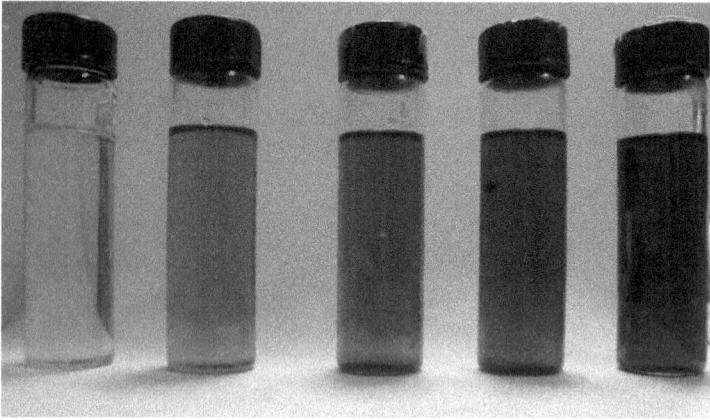

FIGURE 5.5   Photographs of sedimentation of pure oleic acid and oleic acid with different CuO NP mass fractions[16].

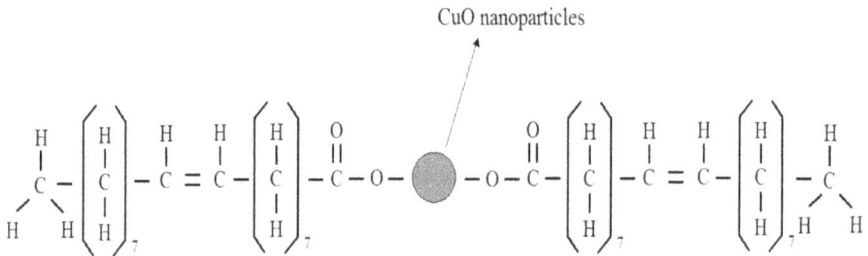

FIGURE 5.6   Chemical structure of CuO-oleic acid nanofluid PCM[16].

The development of defects in nanofluids was demonstrated to be caused by longer periods of high-energy sonication. Ultrasonic temperature is set at 45°C above the melting point of the base liquid to help disseminate the NPs. The CuO NP content of 0.5, 1.0, 1.5, and 2.0% by weight was distributed in molten oleic acid. With the CuO NP concentration in the base fluid increased from 0% to 2%, the change in PCM color could be noticed, as shown in Figure 5.5.

NPs were also put into oleic acid beyond a certain point, causing the nanofluid PCM to turn black because the CuO NP color is black. Figure 5.5 shows improved fluid stability when nanofluid PCMs distributed in CuO-oleic acid havemore constant viscosity. Figure 5.6 shows the chemical structure of oleic acid-CuO NPs. Oleic acid does not dissolve CuO NPs, even if it may be associated with CuO NPs[5].

5.2.3   INFLUENCE OF NPS ON THE THERMAL CONDUCTIVITY

Figure 5.7 presents nanofluid PCMs' heat transfer potential as a function of their NP concentration. The nanofluid PCM's thermal conductivity increased by 0.295, 0.372,

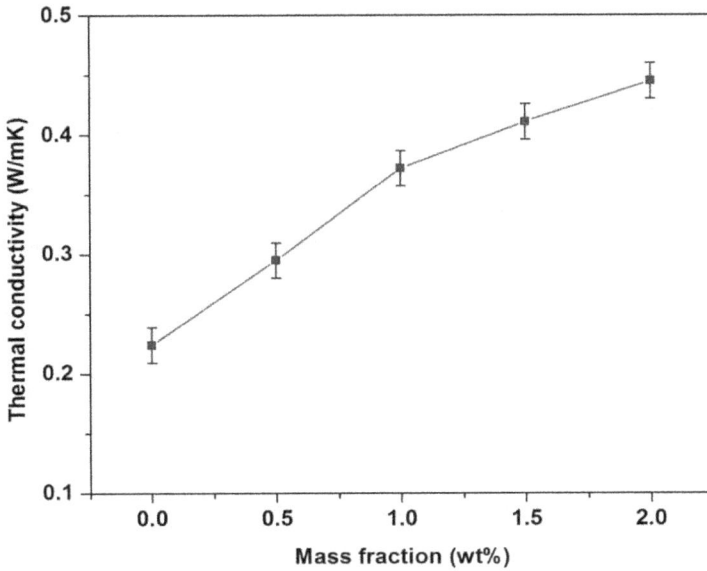

FIGURE 5.7 Thermal conductivity of nanomaterials embedded PCMs[16].

0.415, and 0.445 W/mK correspondingly for CuO NPs of 0.5, 1.0, 1.5 and 2.0 wt%. Also revealed is that the rise in thermal conductivity is linear for lower NP concentrations but nonlinear for larger concentrations. Because the CuO NPs in oleic acid concentrate in greater concentrations, the NPs are more likely to clump together. The agglomeration can dramatically reduce the number of effective scattered NPs, which reduces the increase in heat conductivity.

Thermal conductivity of nanofluid PCMs improved by 31.69, 66.07, 85.26, and 98.66% for 0.5, 1.0, 1.5, and 2.0 wt% of CuO NPs, respectively. CuO-oleic acid has a higher thermal conductivity than oleic acid, which reduces the time it takes to melt and solidify. CuO NPs dispersed into oleic acid have also been used to reduce the melting and solidification times as a nucleating agent.

5.2.4 PHASE CHANGE PROPERTIES OF NANOMATERIALS EMBEDDED PCMS

The DSC test can be used to determine latent heat and phase change temperature in PCMs. The results of each PCM measurement would be significantly different if the PCMs were heated at different rates and with different masses. For the present investigation, DSC readings of the PCMs were carried out at a rate of 5°C per minute.

Figure 5.8 shows the results of the DSC investigations of pure oleic acid and nanomaterial embedded PCMs. It is in this range of temperatures when the solid phase transition of oleic acid and embedded PCMs can be seen as practically flat peaks. There is a maximum temperature of 2°C for solid-liquid oleic acid and a maximum temperature of 6°C for nanomaterial PCMs (melting and solidification point).

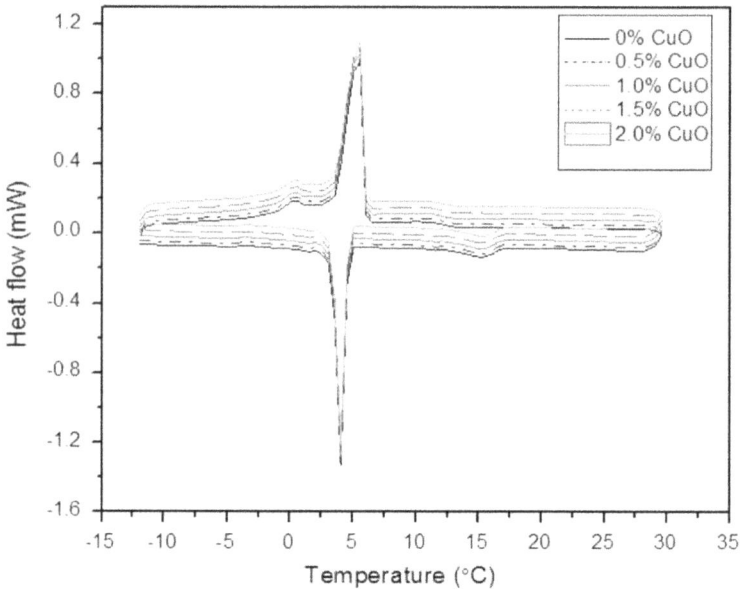

FIGURE 5.8   DSC measurements of CuO-oleic acid nanofluid PCMs[16].

The transition temperatures of nanomaterials in the melting and solidification processes in the embedded PCMs with 0.5wt% and 2.0wt% were determined to be 5.44°C and 5.53°C, respectively. The latent temperatures of the nanomaterials in PCMs with 0.5wt% and 2.0wt% were 136.8 kJ/kg and 133 kJ/kg for melting and 133.4 kJ/kg and 130.4 kJ/kg for solidification, respectively.

Adding CuO NPs resulted in small alterations in both the phase transition temperature and latent temperatures of oleic acid, indicating that there was a strong physical contact between the oleic acid and the CuO NPs. For nanomaterials embedded PCMs with different mass fractions of CuO NPs depending on the number of thermal cycle, similar phase change temperatures and latent heat were reported. The thermal stability of CuO-oleic acid nanofluid PCMs ensures that thermal energy storage systems can perform at a high level.

5.2.5   THERMAL PROPERTIES OF THE NANOMATERIALS EMBEDDED PCMS

The DSC was used to evaluate the thermal properties and specifically the latent thermal storage systems with different mass fractions of CuO NPs and pure nanoparticles embedded in PCMs. The latent heat and phasechange temperature values for base fluids and embedded nanomaterials of CuO NPs are shown in Figures 5.9 and 5.10.

Figure 5.9 shows that melting and solidification had the largest temperature variations of 2.4% and 1.4%, respectively. Figure 5.10 shows the greatest decreases in

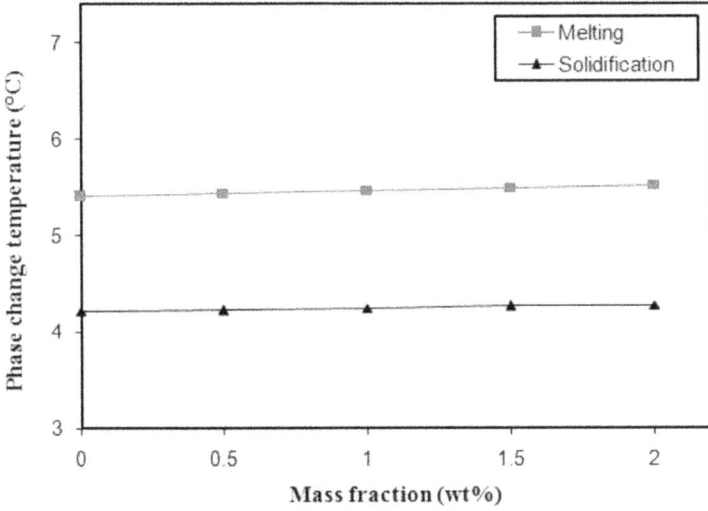

FIGURE 5.9   Variation in PCMs' phase change temperature in relation to mass fraction[16].

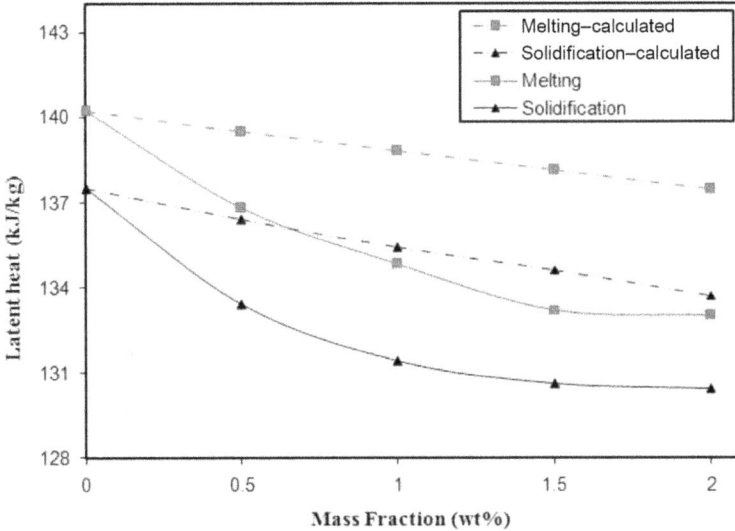

FIGURE 5.10   Variation of PCMs' latent heat relative to mass fraction[16].

latent thermals of embedded nanomaterial PCM melting and solidification of 5.1% and 5.2% respectively. Even though there are some minor variances in the latent heats during both processes, the LTES system will still be able to store and release thermal energy with repeated cycles of operation.

Each nanomaterial PCM's latent heat, as described in Figure 5.10, is less than the latent heat calculated. An increase in the amount of CuO NPs, the latent heat capacity discrepancy, is allied with the increasing heat capacity. This reaction is ascribed to the CuO NPs' interaction with oleic acid. These nanoparticles' thermal properties are influenced by the NP sizes, shapes, and surfaces that exist in the base fluid. As such, a new model is required to analyze the thermal properties of the solid-liquid mixture (nanofluid PCMs)[6].

5.2.6 Thermal Reliability of the Nanomaterials Embedded PCMs

In this section nanofluid PCMs from various CuO NP mass fractions were synthesized for thermal stability after multiple thermal cycles. The temperature fluctuation for phase changes in thermal cycles in relation to 0.5wt% CuO NPs is shown in Figure 5.11. The most change for solidification temperature is −1.65% and for melting temperature −2.59%. This change can be ignored due to the minimal impact it would have on the cool LTES systems.

Figure 5.12 demonstrates the maximum temperature differences from the basis material of 0.87% solidification and 0.92% latent heat melting, respectively. These variations may remain if the energy is stored in latent thermal form. It was found that oleic acid-CuO nanofluid PCM is more thermally stable and reliable than oleic acid, thereby making it the preferred PCM[6].

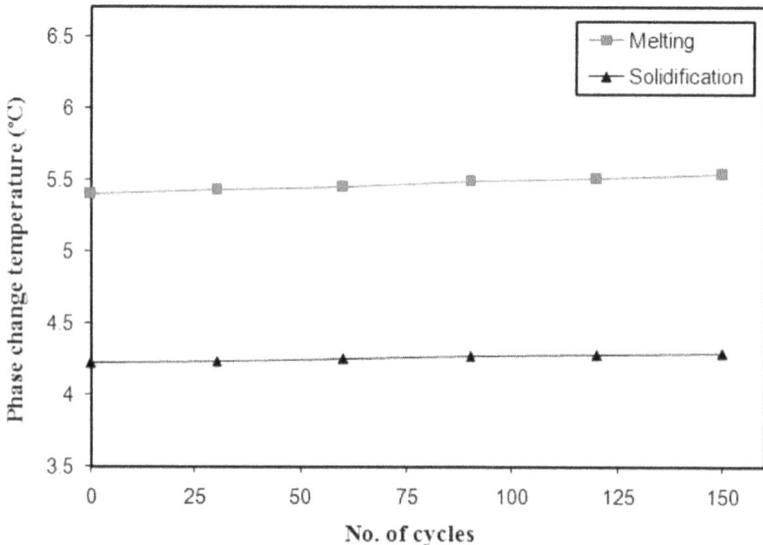

FIGURE 5.11   Thermal cycle-dependent variation in PCM phase change temperature[16].

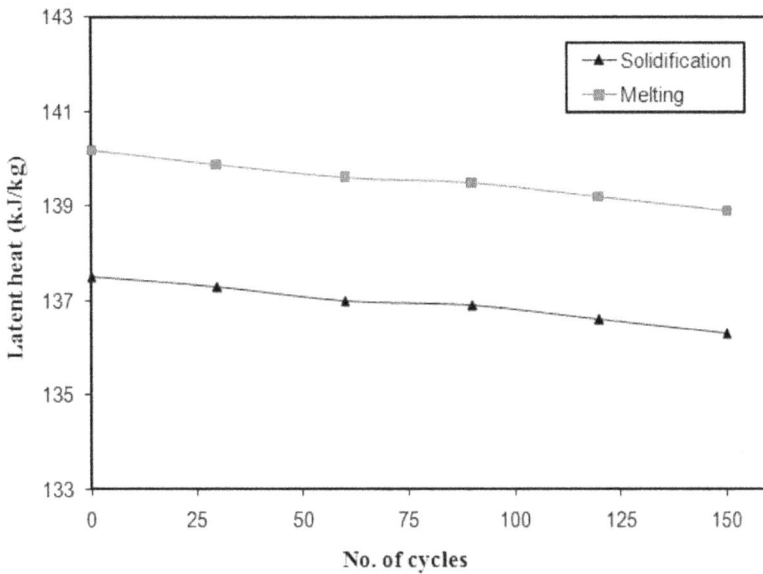

FIGURE 5.12 PCMs' latent heat varies with thermal cycle[16].

5.3 PHASE CHANGE BEHAVIOR OF NANOMATERIALS EMBEDDED PCMs

Thermal and heat transfer capabilities of entrenched PCM nanoparticles were examined during cooling and heating activities. The LTES system will also offer the validation of analytical solutions and the experimental tests concerning PCM freezing and melting features in its pure state and with distributed CuO NPs.

5.3.1 EXPERIMENTAL SETUP

The heat transfer investigations of base PCMs and nanomaterials embedded in PCM during solidification and melting processes were conducted through an experiment. Figure 5.13 depicts the schematic diagram of the experimental setup. The PCM encapsulations are aboded in an insulated TES tank, and a compensating heat transfer tank delivers the heat transfer fluid at a constant temperature during solidification and melting. PCMs were filled in spherical encapsulation of high density polyethylene (HDPE) with a diameter of 70 mm, as shown in Figure 5.14a. The HTF (Heat Transfer Fluid) tank was fitted with the 7 kW cooling unit and the 3000 W heating coil. The cooling unit and heating coil were utilized for cooling or heating the HTF according to the necessary test conditions[7, 16].

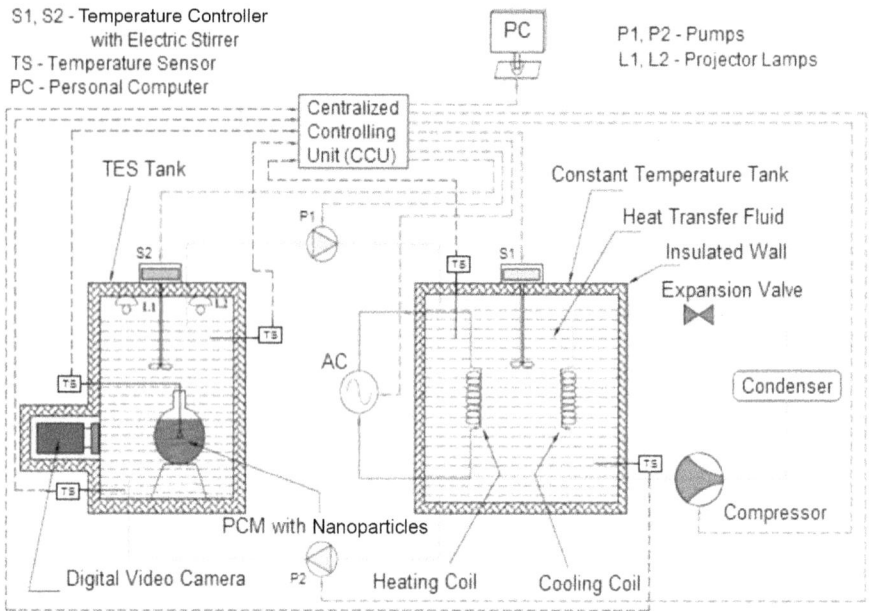

FIGURE 5.13 Experimental setup for testing NPs incorporated in PCM for phase change behavior[16].

As a heat transfer fluid, ethylene glycol (80:20 wt%) was added to water to circulate heat. HTF temperature varies from −20°C to 120°C depending on experimental conditions. External HTF can be circulated through the constant-temperature bath during the experiment to establish isothermal conditions on the encapsulation's wall surface during the testing phase.

The electrical stirrer control was employed to keep the TES tank and HTF bath at the same temperature. Seven thermocouples of J type (Thermo scientific, US) were put within the spherical encapsulation with PCM in the vertical plane at varied radial points in Figure 5.14b. The investigation of transient PCM interfacial positions were from the inner to the center of the spherical encapsulation (i.e. from 0 to 35 mm). The center of the encapsulation was located at the thermocouple 4, while other thermocouples were positioned at different radial positions in Figure 5.14b. The calculated thermocouple uncertainty was ±0.05°C.

In the tanks, both of the temperature controllers were initially set to the same value. The progress in the solidification and melting process of PCMs has been recorded by a digital camera. All sensors are connected to the Centralized Controlling Unit (CCU) to ensure that the LTES system operates properly under various thermal load conditions[8].

During melting, the HTF temperature was maintained at 40°C and delivered by means of the circulating pump to the TES tank. The PCMs were then heated by

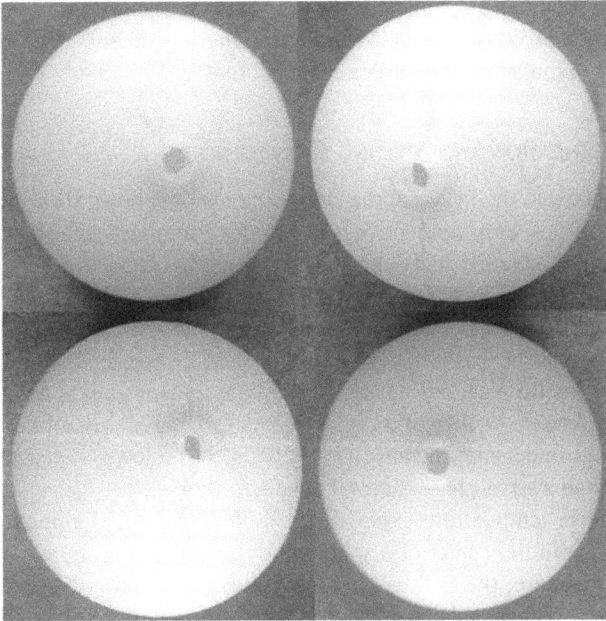

(a)

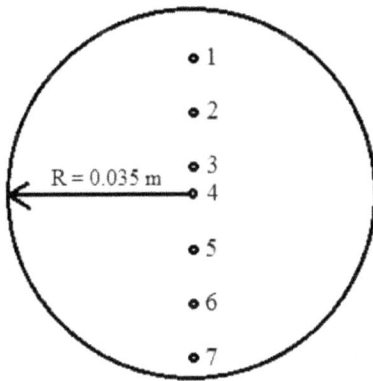

(b)

FIGURE 5.14   (a) Spherical encapsulation, (b) positions of the temperature sensors placed in the encapsulation[16].

the HTF until they were completely melted. This resulted in the repeated melting and solidification processes for PCM and nanomaterials of embedded PCMs for five times in order to make the measurements repetitive.

5.3.2 SOLIDIFICATION AND MELTING CHARACTERISTICS

Heat is transferred to the PCMs at the beginning of the solidification process, and as the process develops, only conductive heat transfer dominates solidification. Because the liquid PCM is situated between the heat transfer surface and the PCM solid region, the solid region gets closer to the heat transfer surface. Because of its higher thermal conductivity, solid PCM solidifies more quickly than liquid PCM. During the melting process, heat is transferred to the PCM by conduction and then natural convection. The solid zone moves away from heat transfer region as the liquid area expands in thickness. As the solid region recedes, the area around the heat transfer region increases in density. Because of this, PCM has a lower liquid-phase thermal conductivity than a solid-phase thermal conductivity.

The change in temperature experienced by oleic acid and embedded NPs during solidification in the center of a spherical encapsulation is shown in Figure 5.15. During the procedure, the PCMs containing oleic acid and NPs are cooled down. Oleic acid and the encapsulated NPs changed from liquid to solid after some time in PCMs. For oleic acid with 0.5% CuO NPs, the solidification process took 1485 s and it took 1635 s to solidify pure oleic acid. System modeling studies of nanofluids including Cu and water for thermal energy storage are in agreement with these findings[9].

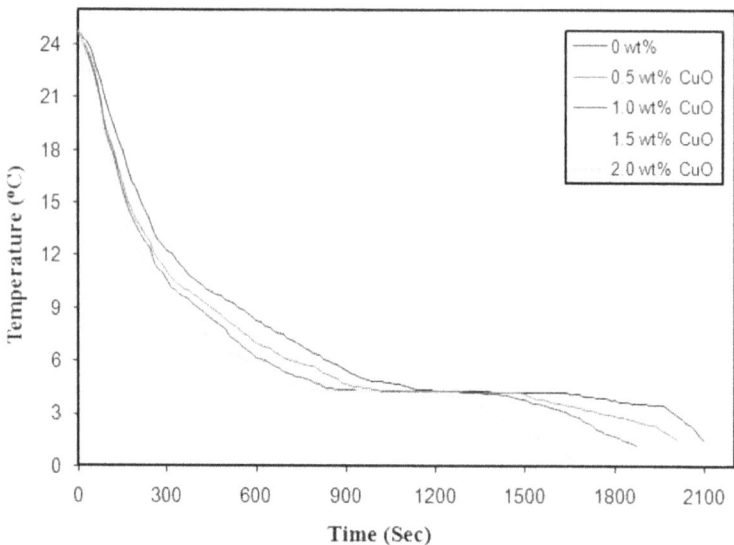

FIGURE 5.15   Solidification curves of nanomaterials embedded PCMs[16].

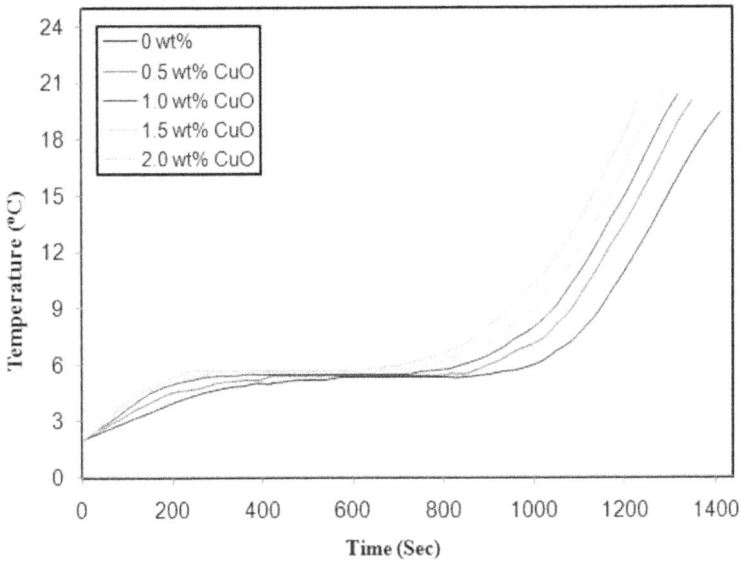

FIGURE 5.16   Melting curves of nanomaterials embedded PCMs[16].

Figure 5.16 shows the temperatures at which pure oleic acid and PCMs begin to melt when heated to 4°C. The temperature of the oleic acid and PCM-incorporated NPs increased as the melting process continued. The temperature remained constant, but the liquid-like PCMs' phases altered with time. In the pure oleic acid, the total melting process time is 840 s, whereas the CuO NPs with mass fractions of 0.5, 1.0, 1.5, and 2.0wt% melt in 780s, 720s, 630s, and 600s, respectively. PCMs and base fluid require different amounts of time to melt and solidify embedded nanomaterials completely. Nanomaterials integrated in PCMs and base fluid have varying thermal conductivity, and the predominant heat transfer mode changes between solidification and melting.

CuO NPs enhance heat transfer in NP embedded PCMs compared to pure oleic acid (NPs). Reductions in the time required for NPs to solidify and melt embedded PCMs indirectly showed the improvement of oleic acid's thermal conductivity, while copper oxide (CuO) particles served as nucleating agents, aiding in faster melting and solidification. Furthermore, the CuO NPs of varying sizes and shapes, such as those present in oleic acid, can enhance the concert of LTES systems.

CuO-oleic acid nanofluid PCM freezing and melting findings are shown in Figures 5.17 and 5.18. The encapsulation's outside wall cooled as heat was transmitted from it to the center. PCMs inserted in the outer container were first frozen, and the process moved toward the center of the container. Figure 5.17 depicts the PCMs implanted in NPs at various stages of freezing, from the outer wall to the center.

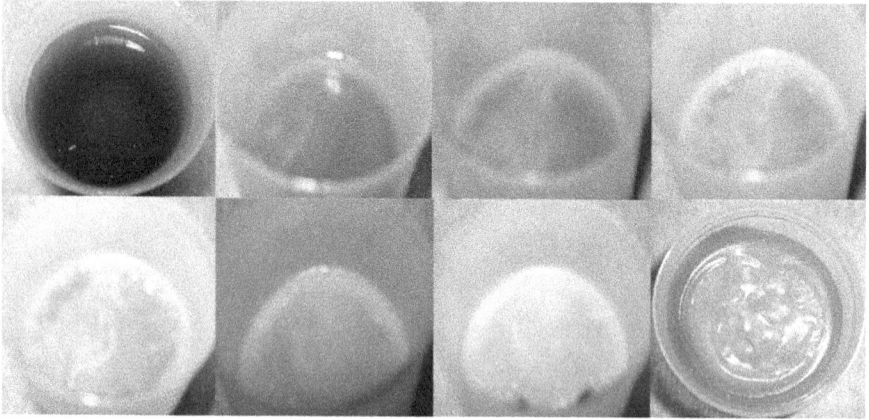

FIGURE 5.17 Photographic view of the progress of the full solidification of NPs embedded PCMs[17].

FIGURE 5.18 Photographic views for progress of complete melting of nanomaterials embedded PCMs[17].

In addition, heat was transferred from the container's exterior wall to the center of the container as it was being heated. Initially, the exterior wall of the encapsulation exhibited melting of the NPs embedded PCMs. The nanomaterials in the regions of incorporated PCMs had a strong condition and were displayed as the ring shape in Figure 5.18. As the time interval continued, the ring's diameter diminished progressively, and then it lost completely. The melting process was apparent to be completed.

5.3.3 SOLIDIFICATION AND MELTING ANALYSES

The variation of the interface between the solid and liquid phases in PCMs and nano-material embedded PCMs over the course of solidification and melting are shown in Figure 5.19. The distance from the center of the spherical capsule to the surface, represented by the y-axis in Figure 5.19a–b, indicates the temporary position of the interface layer for the solid-liquid interface layer, which ranges from 0 to 0.035 m. Additionally, the x-axis describes the amount of time it takes for the PCMs to solidify or melt completely.

As shown in Figure 5.19a, following the testing cycle, the freezing time of the PCMs is greatly reduced, showing that the incorporated nanomaterials could be advantageous. Some well-dispersed NPs were expected to be present even if nano-materials were integrated into PCMs.

These NPs, which are thermally conductive, at the near-wall surface efficiently absorb cold energy and transfer it to nearby PCMs that contain nanomaterials. The presence of NPs in PCMs causes a heterogeneous nucleation process, which facilitates the enlargement of a stable nucleus. Nanomaterials embedded PCMs achieve rapid freezing with faster propagation of the solid-liquid interface; however this leads to the exothermic heat release.

As shown in Figure 5.19b, the heating cycle data suggests that the melting of the nanomaterials embedded PCMs is due to the nanoconvection of the NPs. To accelerate nanomaterial-infused PCMs with quick freezing and melting, it is crucial to have a higher concentration of NPs. The results of the experimental tests and the analytical solution show that increasing the concentration of CuO NPs reduces the time needed for solidification and melting of nanomaterials implanted in PCMs.

An analytical solution has given the predictions that time variations for the complete solidification and melting decreased from 1700 s to 1346 s, and 1100 s to 827 s, and for the corresponding mass fraction of NPs increased from 0.5 wt% to 2.0 wt%. Nanomaterial freezing and melting time periods in PCMs were slashed from 1762 seconds to 1440 seconds and 1149 seconds to 889 seconds, respectively, according to these experimental results. The result shows that the overall melting and solidification processes are deviated by 6.53% and 6.97% from the analytical model solutions.

5.4 WATER-GLYCEROL MIXTURE BASED CuO NANOFLUID PCM

Water-glycerol nanofluid mixes containing CuO nanoparticles are studied in this section for their thermal energy storage capabilities. The mixture was presented to be 80% water and 20% glycerol (by weight). The water-glycerol mixture was mixed with CuO NPs of mass fractions of 0.1, 0.3, 0.5, 0.8, and 1.0wt% separately.

5.4.1 CHARACTERIZATION OF CuO NPs

It is shown in Figure 5.20 that the CuO particles are within the 12 and 50 nm size range, and the NPs appear to be flake-like in structure. These nanoflakes could prove useful in the water-glycerol mixture to improve its thermal conductivity. As can be

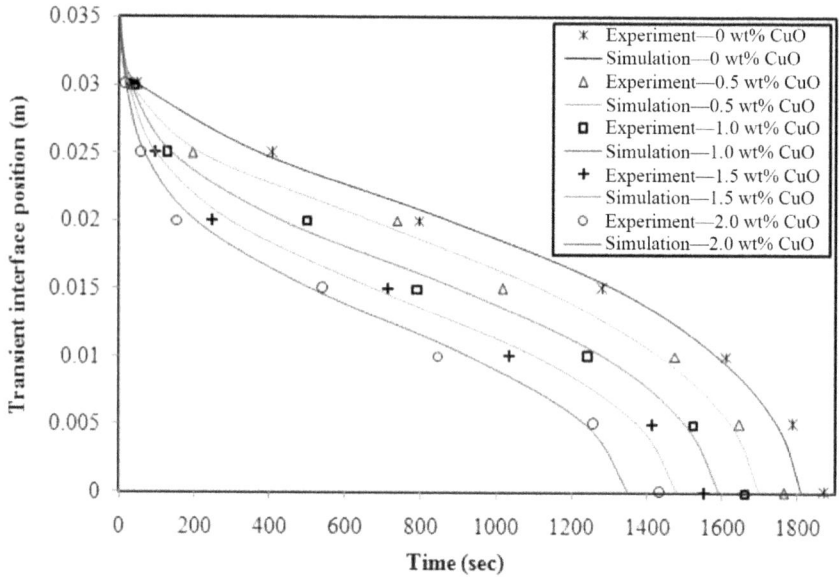

(a)

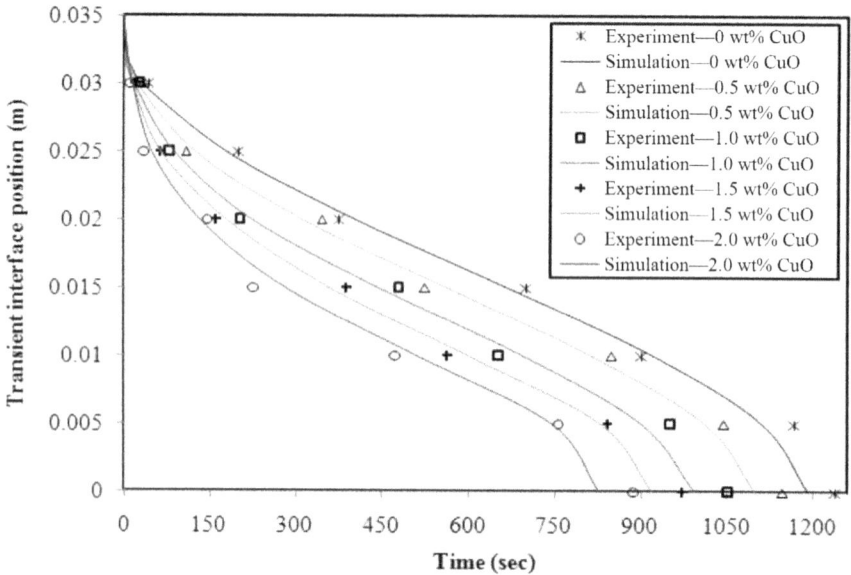

(b)

FIGURE 5.19 Comparison between experimental and analytical results for (a) solidification, (b) melting[17].

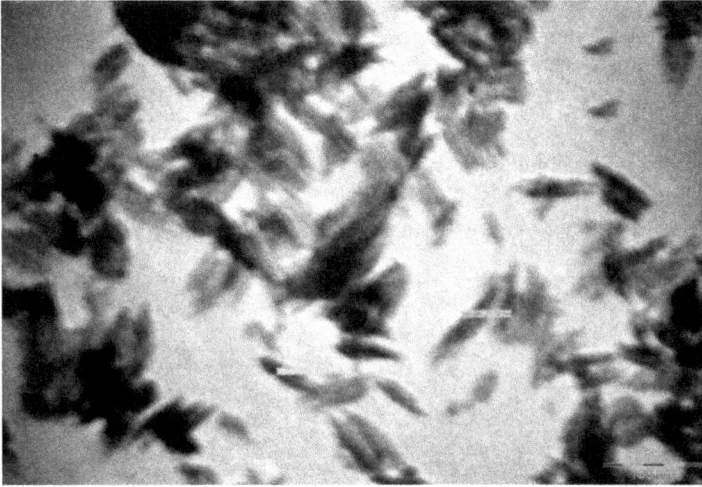

FIGURE 5.20   TEM image of CuO nanoflakes[17].

100 nm   EHT = 5.00 kV   Signal A = InLens   Gun Vacuum = 3.32e-009 mbar
         WD = 5.3 mm   Mag = 150.15 K X

FIGURE 5.21   SEM image of CuO nanoflakes[17].

seen in Figure 5.21, the CuO particles produced by the water precipitation process have a flake-like structure. This flake-like structure would offer a highest surface area of heat transfer contact with the base fluid than the other types with rectangular and spherical structures.

The XRD pattern of the CuO nanoflakes reveals a single crystal structure, as demonstrated in Figure 5.22. The lattice parameters, a = 4.79 Å, b = 3.42 Å, and c = 5.29 Å, were found. The XRD pattern has no impurity peaks present. The observed XRD pattern matched JCPDS data rather well (file No. 45–0937).

A single peak in the particle size analyzer result indicates that the CuO NPs are uniformly dispersed in the base fluid, and that there is no clumping in the nanomaterials included in the PCMs. CuO particle sizes were measured by the PSA and ranged from 14 nm to 70 nm, with an average of 17 nm (Figure 5.23).

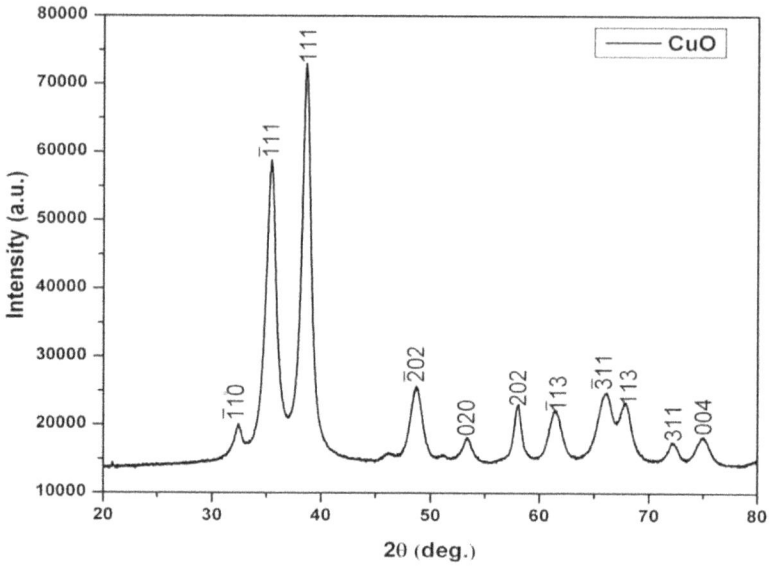

FIGURE 5.22   XRD pattern of CuO NPs[17].

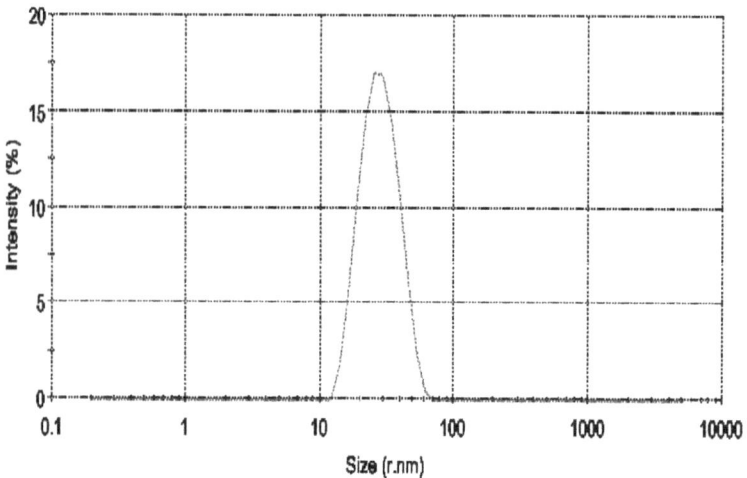

FIGURE 5.23   Particle size distribution of CuO NPs in water-glycerol mixture[17].

5.4.2 PREPARATION OF NANOMATERIALS EMBEDDED PCMs

The water and glycerol mass fractions were combined in a weight ratio of 80:20 to make up the base fluid for the creation of PCMs that are embedded in nanomaterials. Table 5.2 presents the thermophysical characteristics of the water/glycerol mixture. Using a two-step process, different CuO NP mass fractions of 0.1, 0.3, 0.5, 0.8, and 1.0 wt% were distributed into the base fluid (i.e. water-glycerol combination). There must be no chemical reaction between the base fluid and the NPs while creating PCM-containing nanomaterials.

Ultrasonic dispersal technique was utilized to ensure that the base fluid would have consistent dispersion of NPs and hence low or no agglomeration of NPs. Ultrasonic vibrators (frequencies of 40 kHz) were used to measure the residing times of nanomaterials with varying mass fractions. The materials were maintained for various lengths of time. Because the NPs in PCMs are unstable, prolonged residing times lead to defects in the embedded nanomaterials.

It was decided to increase the ultrasonic temperature to 45°C in order to improve the dispersion of the NPs. The water-glycerol solution contained CuO NPs at concentrations of 0.1, 0.3, 0.5, 0.8, and 1.0wt%. In the sedimentation image shown in Figure 5.24, CuO nanofluid PCM dispersion stability was found to be excellent.

TABLE 5.2
Thermophysical Properties of the Water-Glycerol Mixture[17]

S.No	Properties of the PCM	Values
1.	Melting temperature (°C)	(3–4)
2.	Density (kg/m³)	1058
3.	Latent heat (kJ/kg)	306.42
4.	Thermal conductivity (W/m K)	0.534
5.	Viscosity (mPa-s)	1.731

FIGURE 5.24 Sedimentation photographs of base fluid and nanomaterials embedded PCMs[17].

5.4.3 INFLUENCE OF NPs ON THE THERMAL CONDUCTIVITY

Figure 5.25 shows the results of testing the thermal conductivity of the PCM nanomaterials and the base fluid. A thermal conductivity increase of nanomaterials contained PCMs was seen with CuO NPs at a 0.1, 0.3, 0.5, 0.8, and 1.0wt% concentration, with thermal conductivities of 0.594, 0.641, 0.692, 0.741, and 0.791 W/mK, respectively. This is a result of clumping, and it could have a significant impact on the LTES system's heat transfer properties. The thermal conductivity of the PCMs in nanomaterials increased by 5.32, 13.65, 22.69, 31.38, and 40.24% compared to the base fluid for 0.1, 0.3, 0.5, 0.8, and 1.0 wt% correspondingly.

5.4.4 EFFECT OF NPs AND TEMPERATURE ON THE VISCOSITY

Figures 5.26 and 5.27 depict the relationship between viscosity and the concentration of NPs in the base fluid and the operating temperature of the NPs embedded PCMs. Figure 5.26 shows that the viscosity increases linearly as the concentration of NPs increases. NPs are shaped differently and have a variety of sizes and pH values, leading to a fluctuation in concentration.

The addition of NPs to the base fluid will not negatively impact the operation of the LTES system, as viscosity is expected to increase. Different from LTES, nanofluids flow in either a counter or parallel direction with increased viscosity in the dynamic flow system, altering the heat transfer rate between nanofluids and HTF by increasing resistance to fluid flow. CuO NPs may operate as a nucleating agent

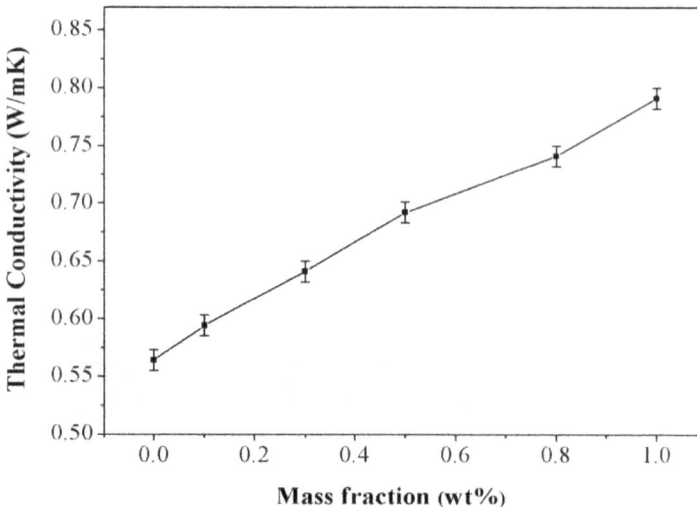

FIGURE 5.25   Effect of CuO NP mass fractions on thermal conductivity[17].

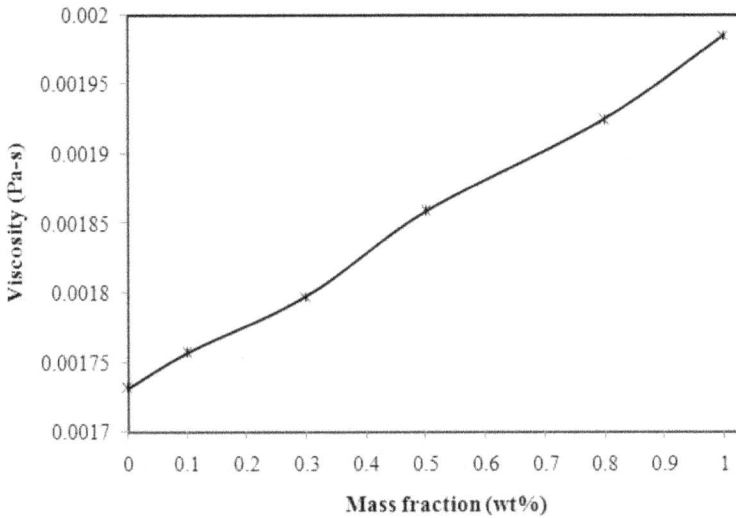

FIGURE 5.26 Effect of mass fractions of CuO NPs on viscosity[17].

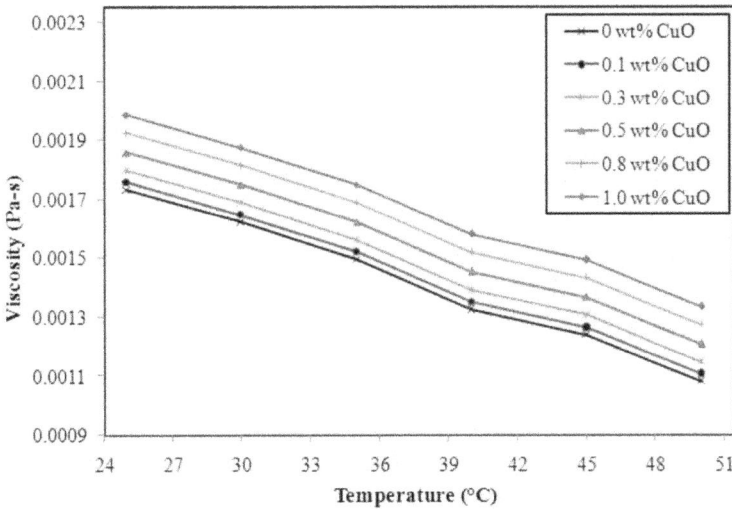

FIGURE 5.27 Variation of viscosity with respect to temperature[17].

in the base fluids, reducing the supercooling impact of the water-glycerol mixture. Figure 5.27 shows that the viscosity of PCM decreases linearly as the working temperature increases, which is completely different from the viscosity variation as a function of NP mass fraction.

5.4.5 THERMOGRAVIMETRIC (TG) ANALYSIS

Figure 5.28 shows the TG curves of the base fluid and nanofluid PCMs containing 0.5 and 1.0 wt% CuO NPs. In the temperature range of 40°C to 90°C, the base fluid and PCM-incorporated nanoparticles decline significantly. The agitation of the water-glycerol mixture in the base fluid and the incorporation of nanomaterials in the PCMs are the primary causes.

Nanofluids with 1.0 wt% CuO NPs lose less weight than those with 0.5 wt% CuO NPs. CuO NPs incorporated in the base fluid had much better thermal stability than the base fluid alone. The working temperature of the nanomaterials embedded PCMs would be in the range of −8°C to 34°C in terms of cooling thermal energy storage. As a result, the developed PCMs for the LTES system can function well in this temperature range[8–13].

5.5 PHASE CHANGE PROPERTIES OF NANOMATERIALS EMBEDDED PCMs

5.5.1 THERMAL PROPERTIES OF NANOMATERIALS EMBEDDED PCMs

Figure 5.29 shows the results of DSC measurements performed to establish the PCMs' thermal properties. Figure 5.30 depicts the PCM phase change temperatures, which exhibit almost comparable values for the base fluid and nanomaterials embedded PCMs for different CuO NP mass fractions. Figure 5.30 shows the largest deviations from the base fluid (without CuO NP) for melting and solidification phase

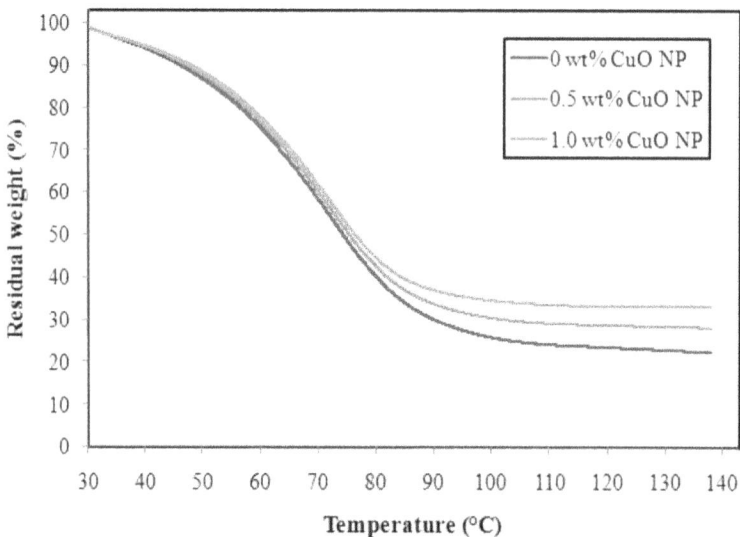

FIGURE 5.28 TG curve for base fluid and nanomaterials embedded PCMs[17].

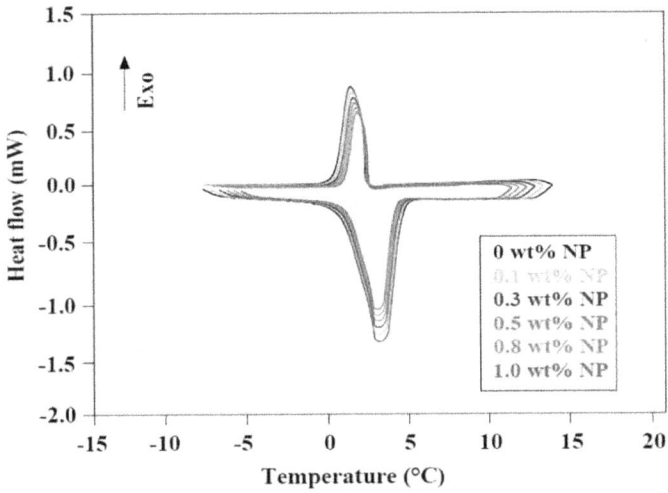

FIGURE 5.29 DSC measurements of water-glycerol mixture based CuO nanofluid PCMs[17].

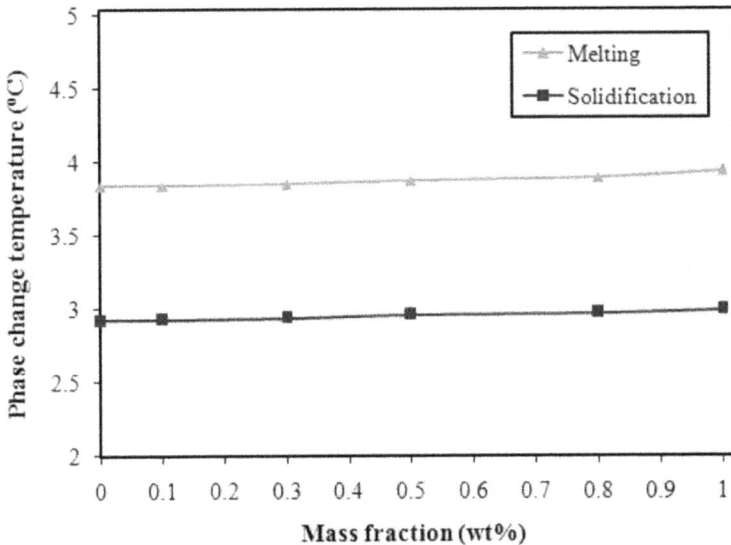

FIGURE 5.30 Phase change temperature varies depending on mass fraction[17].

transition temperatures when nanomaterials embedded PCMs (1.0wt%) are used. Similarly, the maximum deviations from the base fluid for the melting and solidification latent temperatures of embedded nanomaterials in PCMs are 1.85% and 1.57%, respectively (Figure 5.31).

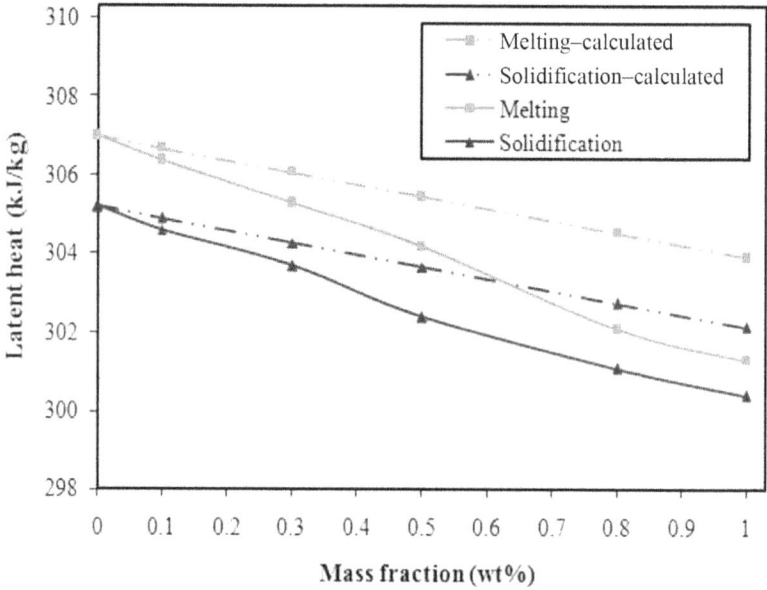

FIGURE 5.31   Changes in latent heat as a function of mass fraction[17].

Small fluctuations in latent heat can be tolerated in the LTES systems due to their rapid energy storage and release rates. Furthermore, these little variations would have no negative impact on the PCMs' ability to store and release energy. The latent temperatures of NPs embedded PCMs for melting and solidification were determined using the mass and latent heat of the base fluid[14].

5.5.2   Thermal Reliability of Nanomaterials Embedded PCMs

PCM thermal reliability is critical to the LTES system's long-term performance. Nanomaterials embedded PCM with 1.0wt% CuO NPs were used to investigate the thermal reliability properties. Thermal cycles were used to determine the PCMs' latent heat and phase change temperatures implanted with nanomaterials.

Figure 5.32 shows that the maximum phase change temperature variations for solidification and melting of nanomaterials embedded PCMs were −2.21 and −2.40%. Figure 5.33 shows the maximum latent heat temperature variations for solidification and melting at 0.91% and 0.76%, respectively[15].

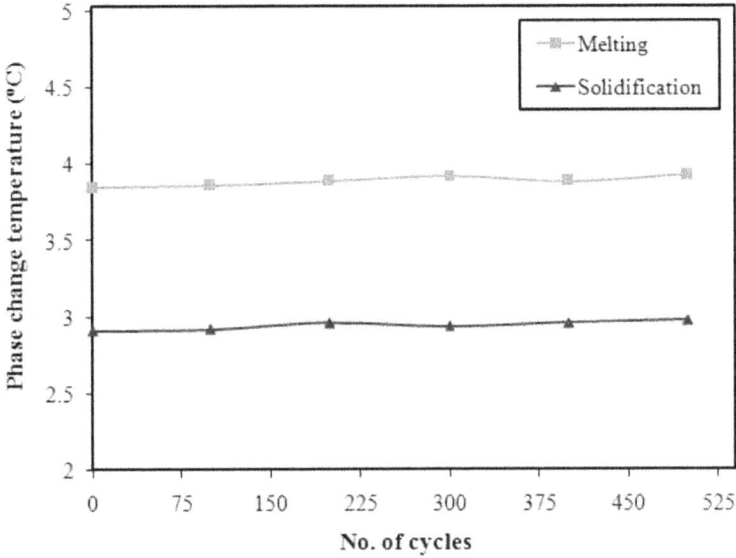

FIGURE 5.32 The number of cycles has an effect on the phase change temperature[17].

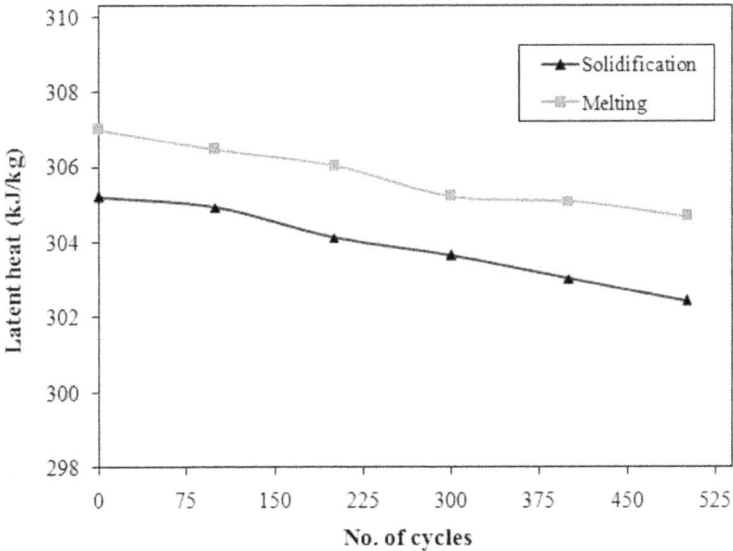

FIGURE 5.33 Latent heat variation according to the number of cycles[17].

5.6 SUMMARY

CuO-oleic acid nanofluid PCM and water-glycerol combination based CuO nanofluid PCM were investigated experimentally. The thermal properties of CuO NPs distributed in oleic acid and a water-glycerol mixture using a two-step process and analyzed experimentally were investigated. CuO NPs were found to be evenly distributed and diffused in both base fluids, according to the particle size analyzer results.

After a maximum number of thermal cycle PCM operations, the variances in latent heat and phase change temperatures were observed to be insufficient, showing that both the nanomaterials embedded PCMs had improved thermal dependability over their respective base fluids. CuO nanofluid PCM based on water-glycerol mixture increased in viscosity as NP concentration in base fluid increased. The viscosity of nanomaterials embedded PCMs, on the other hand, reduced significantly when the operating temperature of the nanomaterials embedded PCMs climbed. The thermal conductivity of CuO nanofluid PCMs based on a water-glycerol mixture increased by 5.32, 13.65, 22.69, 31.38, and 40.24% in comparison to the base fluid. The thermal conductivity improvement of CuO-oleic acid nanofluid PCMs for 0.5, 1.0, 1.5, and 2.0wt% was 31.69, 66.07, 85.26, and 98.66%, respectively, over the base fluid.

CuO-oleic acid nanofluid PCMs with 0.5, 1.0, 1.5, and 2.0wt% reduced complete solidification durations by 10.71, 16.07, 19.64, and 27.67%, respectively, as compared to pure oleic acid. When compared to pure oleic acid, CuO-oleic acid nanofluid PCMs with 0.5, 1.0, 1.5, and 2.0wt% reduced complete solidification times by 10.71, 16.07, 19.64, and 27.67%, respectively. The analytical solutions were also confirmed with the experimental results for both the solidification and melting processes. As a result of their increased thermal characteristics, thermal reliabilities, and heat transfer performance, the nanomaterials embedded PCMs explored in this chapter could be recommended for cool thermal storage.

REFERENCES

[1] Gunther, E., Hiebler, S., Mehling, H. & Redlich, R., 'Enthalpy of phase change materials as a function of temperature: Required accuracy and suitable measurement methods', *International Journal of Thermophysics*, vol. 30, pp. 1257–1269, 2009.

[2] He, Q., Wang, S., Tang, M. & Liu, Y., 'Experimental study on thermophysical properties of nanofluids as phase-change material (PCM) in low temperature cool storage', *Energy Conversion and Management*, vol. 64, pp. 199–205, 2012.

[3] Ji, M., Kim, J. H., Ryu, C. H. & Lee, Y. I., 'Synthesis of self-modified black BaTiO3-x nanoparticles and effect of oxygen vacancy for the expansion of piezocatalytic application', *Nano Energy*, p. 106993, 2022.

[4] Maghrabie, H. M., Elsaid, K., Sayed, E. T., Radwan, A., Abo-Khalil, A. G., Rezk, H., . . . Olabi, A. G., 'Phase change materials based on nanoparticles for enhancing the performance of solar photovoltaic panels: A review', *Journal of Energy Storage*, vol. 48, p. 103937, 2022.

[5] Li, C. H. & Peterson, G. P., 'Experimental investigation of temperature and volume fraction variations on the effective thermal conductivity of nanoparticle suspensions (nanofluids)', *Journal of Applied Physics*, vol. 99, no. 8, pp. 1–8, 2006.

[6] Murshed, S. M. S., Leong, K. C. & Yang, C., 'Enhanced thermal conductivity of TiO$_2$-water based nanofluids', *International Journal of Thermal Sciences*, vol. 44, pp. 367–373, 2005.

[7] Na, Y. S., Kihm, K. D. & Lee, J. S., 'Opposite Re$_D$-dependencies of nanofluid (Al$_2$O$_3$) thermal conductivities between heating and cooling modes', *Applied Physics Letters*, vol. 101, no. 8, p. 083111, 2012.

[8] O'Connell, M. J., Bachilo, S. M., Huffman, C. B., Moore, V. C., Strano, M. S., Haroz, E. H., Rialon, K. L., Boul, P. J., Noon, W. H., Kittrell, C., Ma, J., Hauge, R. H., Weisman, R. B. & Smalley, R. E., 'Band gap fluorescence from individual single-walled carbon nanotubes', *Science*, vol. 297, pp. 593–596, 2002.

[9] Zhang, T., Juan, S., Yi, W. & Chen, Z., 'A novel form-stable phase change material based on halloysite nanotube for thermal energy storage', *Journal of Energy Storage*, vol. 45, p. 103703, 2022.

[10] Das, P., Kundu, R., Kar, S. P. & Sarangi, R. K., 'Fabrication of composite phase change material: A critical review', *Advancement in Materials, Manufacturing and Energy Engineering*, vol. 2, 97–106, 2022.

[11] Hadavimoghaddam, F., Atashrouz, S., Rezaei, F., Munir, M. T., Hemmati-Sarapardeh, A. & Mohaddespour, A., 'Modeling thermal conductivity of nanofluids using advanced correlative approaches: Group method of data handling and gene expression programming', *International Communications in Heat and Mass Transfer*, vol. 131, p. 105818, 2022.

[12] Wu, R., Ma, Z., Gu, Z. & Yang, Y., 'Preparation and characterization of CuO nanoparticles with different morphology through a simple quick-precipitation method in DMAC-water mixed solvent', *Journal of Alloys and Compounds*, vol. 504, pp. 45–49, 2010.

[13] Xie, H., Chen, L. & Wu, Q., 'Measurements of the viscosity of suspensions (nanofluids) containing nanosized Al$_2$O$_3$ particles', *High Temperatures-High Pressures*, vol. 37, pp. 127–135, 2008.

[14] Scholes, C. A., 'Applications of membranes with nanofluids and challenges on industrialization', In: *Nanofluids and Mass Transfer*, Elsevier, pp. 385–398, 2022.

[15] Wu, R., Ma, Z., Gu, Z. & Yang, Y., 'Preparation and characterization of CuO nanoparticles with different morphology through a simple quick-precipitation method in DMAC-water mixed solvent', *Journal of Alloys and Compounds*, vol. 504, pp. 45–49, 2010.

[16] Harikrishnan, S. & Kalaiselvam, S., 'Preparation and thermal characteristics of CuO-oleic acid nanofluids as a phase change material', *Thermochimica Acta*, vol. 533, pp. 46–55, 2012.

[17] Harikrishnan, S., Magesh, S. & Kalaiselvam, S., 'Preparation and thermal energy storage behaviour of stearic acid-TiO2 nanofluids as a phase change material for solar heating systems', *Thermochimica Acta*, vol. 565, pp. 137–145, 2013.

6 NEPCMs for Heating Applications

CONTENTS

6.1 INTRODUCTION

For preparing NPs embedded PCMs, a two-step approach was used. An investigation of the thermal properties of two pure PCMs and nanomaterial-incorporated PCMs was carried out in terms of mass fraction and number of thermal cycles. Laser flash analyzer tests were employed to confirm that the inclusion of NPs had increased the thermal conductivity of the PCMs. The addition of NPs to the base fluids also caused

DOI: 10.1201/9781003163633-6

an increase in viscosity, which was studied to determine if this helped the heat transfer performance of the NPs in the PCMs during heating and cooling. Analytically and experimentally, the solidification and melting characteristics of the nanomaterials embedded PCMs were examined. The studies for the two PCMs described previously are discussed in detail in the following sections.

6.2 STEARIC ACID-TiO$_2$ NANOFLUID PCM

The synthesis and characterization of stearic acid-TiO$_2$ nanofluid PCM are discussed in this section. Thermal and heat transfer performance was also examined for the NPs embedded PCMs by testing their consistency in melting and solidifying. During cooling and heating processes, the transient interface locations of NPs embedded PCMs were investigated analytically.

6.2.1 CHARACTERIZATION OF TiO$_2$ NPs

Figure 6.1 shows a TEM image of TiO$_2$ particle sizes ranging from 17 to 85 nm. Despite the fact that particle sizes differed, particle dispersion was consistent. The larger-sized particles are found on the left and right sides of the image, together with the smaller-sized particles. This small agglomeration will not significantly improve thermal conductivity and can be managed with an ultrasonic vibrator and a suitable surfactant. The sol-gel approach produced TiO$_2$ NPs that were practically spherical in shape.

Figure 6.2 depicts the dispersion of NPs in the base fluid, showing the particle size distribution of TiO$_2$. The base fluid particle size distribution shows a single peak, indicating homogeneous dispersion of TiO$_2$ NPs and the absence of agglomeration. The TiO$_2$ NPs in the base fluid ranged in diameter from 13 nm to 65 nm, with the average particle size of 23 nm, as shown in Figure 6.2.

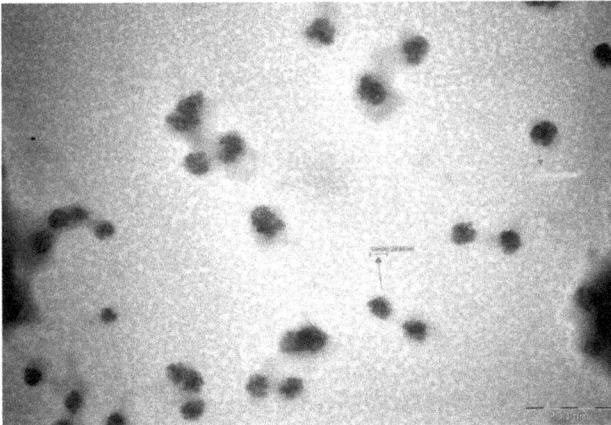

FIGURE 6.1   TEM image of TiO$_2$ NPs[1].

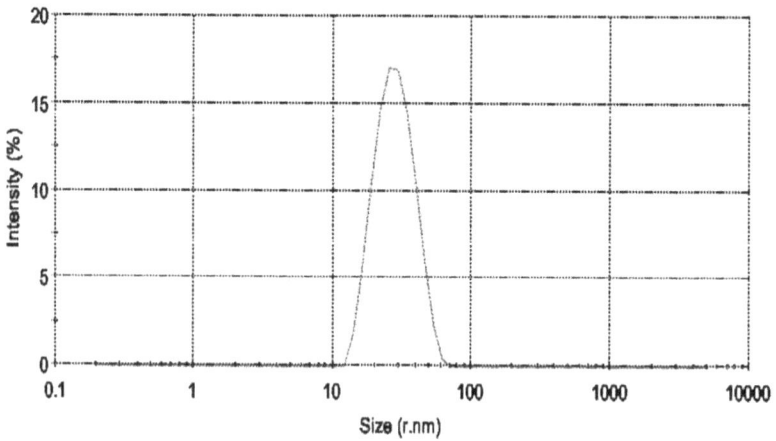

FIGURE 6.2   Particle size distributions of the TiO_2 NPs[1].

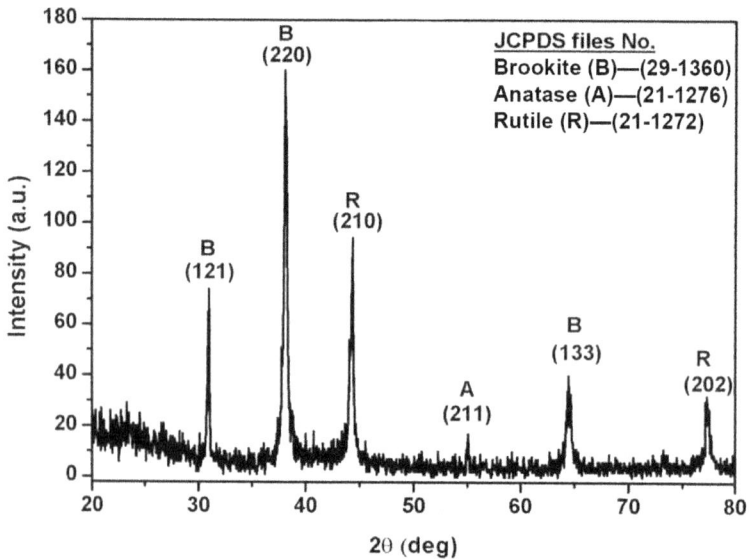

FIGURE 6.3   XRD Pattern of TiO_2 NPs[1].

Images of TiO_2 NPs created by sol-gel synthesis are shown in Figure 6.3 (XRD). In this XRD image, the three phases of TiO_2 particles can be seen clearly (Brookite, Anatase, and Rutile). The annealing temperature of 600°C is to blame for the three phases. The average crystallite sizes were computed using Debye-relation Scherrer's and were found to be between 26 and 29 nm. The XRD pattern also lacked any impurity peaks. The JCPDS files 29–1360, 21–1276, and 21–1272 agree with Brookite, Anatase, and Rutile TiO_2 NPs, respectively.

6.2.2 Preparation of Nanomaterials Embedded PCMs

Thermofisher Laboratories Scientific Private Limited in India supplied the stearic acid ($CH_3(CH_2)16COOH$). In Table 6.1, stearic acid's thermophysical characteristics are listed.

Preparation of NPs integrated PCMs is critical for increasing the system's heat transfer characteristics, since nano-sized particles have a notable positive impact on the base material. Certain characteristics, such as steady dispersion, low particle agglomeration, and no chemical change in the fluid, must be considered during the preparation of nanomaterials embedded PCMs. For consistent dispersion of NPs into stearic acid, sodium dodecylbenzenesulfonate (SDBS) was chosen as the capping agent. SDBS was acquired from Sigma-Aldrich in the USA.

Furthermore, preparation activities were carried out utilizing an ultrasonic vibrator at a frequency of 40 kHz to increase particle dispersion and minimize aggregation of the nanofluid PCMs. Longer residence in the vibrator would allow blemishes to appear in nanofluid PCMs. The vibrator temperature was sustained at a temperature higher than the melting point of the stearic acid during the preparation procedure to maintain the stearic acid in a liquid state. Figure 6.4 depicts an image representation of the preparation of nanofluid PCMs.

6.2.3 Chemical Stability of Nanomaterials Embedded PCMs

Specific interactions in NPs embedded PCM can be revealed using FTIR spectroscopy. Figure 6.5 shows the FTIR absorption spectra of the distinctive peaks for stearic acid and NPs embedded PCMs. The pure stearic acid spectrum displays adsorption peaks at wave numbers of 2918.15 cm^{-1} and 2849.72 cm^{-1}, which correspond to the absorption band of aliphatic C-H vibration caused by stretching vibration of the O-H group.

Stearic acid's aliphatic chain has a peak at 1464.24 cm^{-1}, 1389 cm^{-1} represents C-H and C-C bending, and 720.10 cm^{-1} and 688.74 cm^{-1} reflect rocking vibration and bending. There were no new peaks seen in the FTIR measurements for stearic acid and nanomaterials embedded PCMs, indicating that there were only strong physical relations among stearic acid and TiO_2 NPs and denial chemical reaction.

TABLE 6.1
Thermophysical Properties of the Palmitic Acid

S.No	Properties of the PCM	Values
1.	Molar mass	256.42
2.	Melting temperature (°C)	60–62
3.	Density (kg/m^3)	852
4.	Latent heat (kJ/kg)	140.27
5.	Specific heat capacity (kJ/kg K)	1.792
6.	Thermal conductivity (W/m K)	0.174

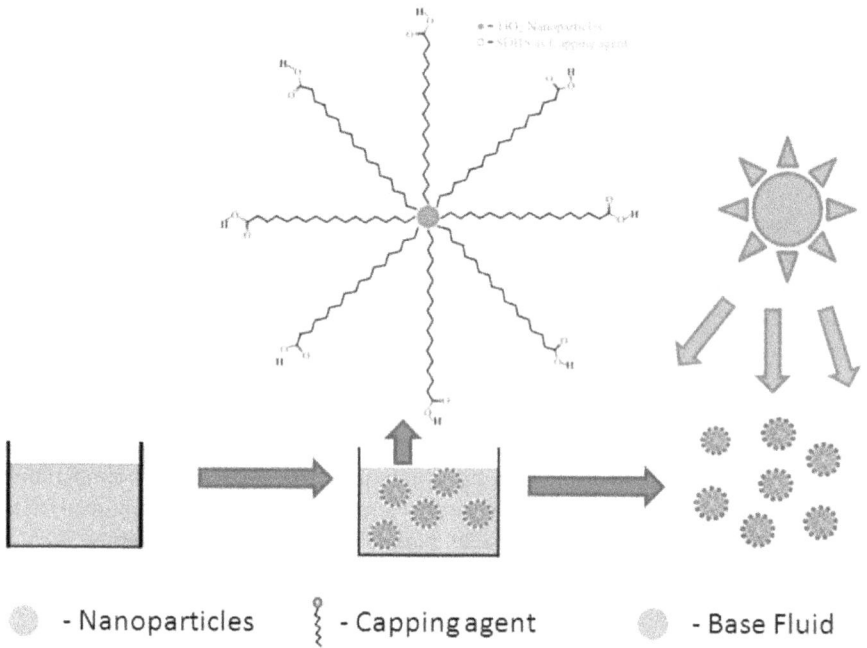

FIGURE 6.4 Pictorial representation for preparation of stearic acid-TiO$_2$ nanofluid PCMs[1]

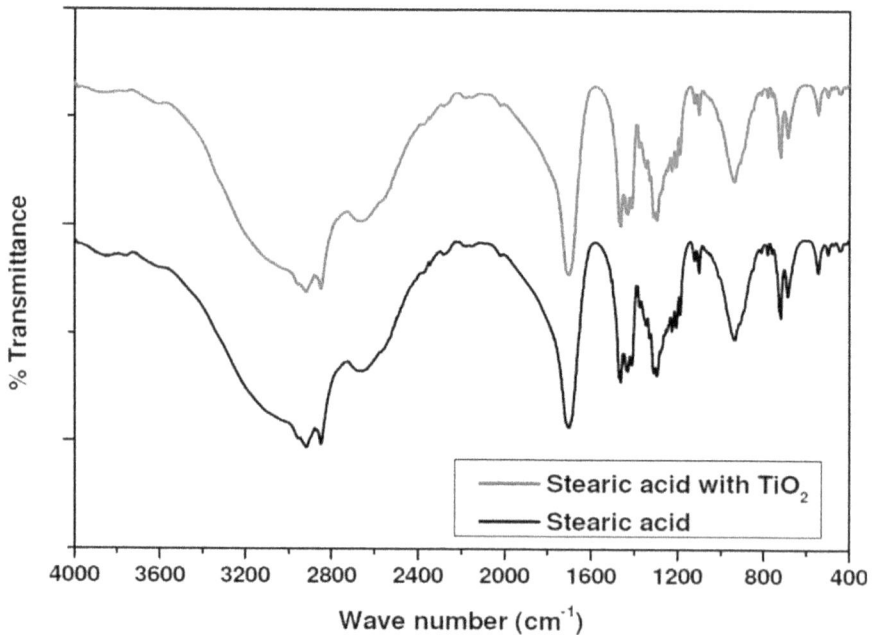

FIGURE 6.5 FTIR spectra of stearic acid and nanofluid PCM with 0.3wt% TiO$_2$ NPs[1]

6.2.4 INFLUENCE OF NPs ON THE THERMAL CONDUCTIVITY

The temperature differential between HTF and PCM and the thermal conductivity of PCM are the most significant factors in the LTES system's melting and solidification processes. PCM energy storage and release times would be significantly improved with thermal conductivity enhancement. Because TiO_2 NPs were mixed with stearic acid, the thermal conductivity of the nanofluid PCMs should be privileged than that of pure stearic acid. The thermal conductivity of nanofluid PCMs as a function of the mass fraction of NPs is shown in Figure 6.6.

The thermal conductivity of nanomaterials embedded PCMs increased by 0.233, 0.263, 0.281, 0.297, 0.312, and 0.324 W/mK for 0.05, 0.1, 0.15, 0.2, 0.25, and 0.3wt% TiO_2 NPs, respectively. In addition, the thermal conductivity of the PCMs increases as the mass fraction of NPs increases. NPs mass fraction has a direct effect on thermal conductivity of NPs implanted PCMs. A higher concentration of NPs could result in a higher thermal conductivity, although the advantage may not be linear. This could be owing to the fact that increased TiO_2 NP concentrations in the base material cause NP agglomeration. The mass fraction of the NPs should be adjusted to improve the thermal conductivity of the PCMs.

Thermal conductivity of TiO_2-Stearic acid was improved by 21.05, 36.84, 47.89, 56.31, 63.15, and 70.53% for 0.05, 0.1, 0.15, 0.2, 0.25, and 0.3wt% NPs when compared to stearic acid. As a result of the increased thermal conductivity, nanofluid PCMs have shorter melting and solidification durations than pure stearic acid. The TiO_2 NPs in stearic acid diffused as a nucleating agent, facilitating with the melting and solidification of the PCM and lowering the supercooling effect.

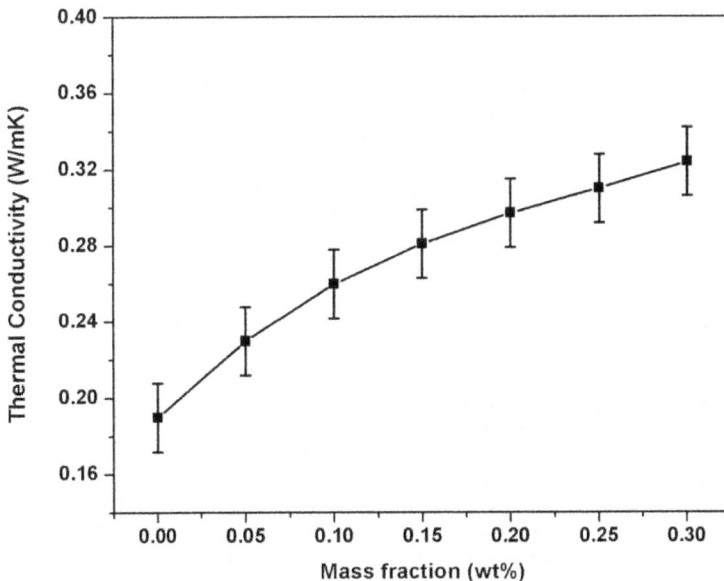

FIGURE 6.6 Influence of NPs on the thermal conductivity[1]

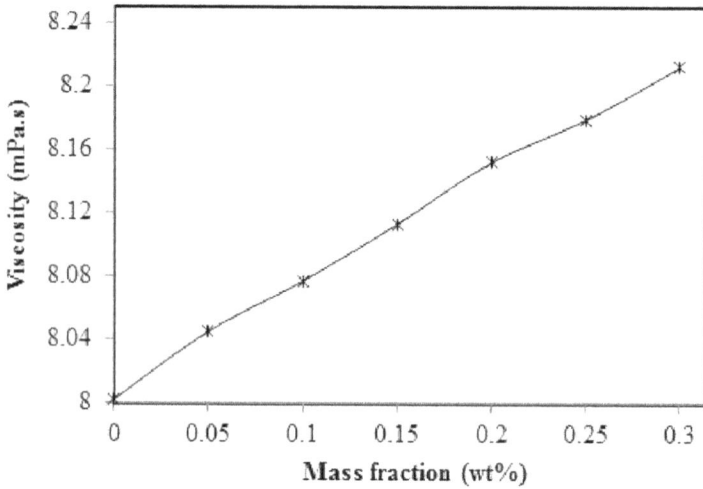

FIGURE 6.7   Effect of NPs on the viscosity[1]

6.2.5   Effect of NPs on the Viscosity

Figure 6.7 shows the change in viscosity as NPs are added to the base material. At 65°C, the viscosities of the basic material and incorporated NPs PCMs were determined. The findings show that as the concentration of NPs rises, so does the viscosity of the NPs embedded PCMs. The viscosity improvement was 0.53, 0.94, 1.38, 1.87, 2.21, and 2.62% for nanomaterials embedded PCMs with 0.05, 0.1, 0.15, 0.2, 0.25, and 0.3wt%, respectively. All the variables affecting viscosity, such as particle size and shape, the presence of NPs, and the pH of the solution, are essential to increase the viscosity.

The viscosity of NPs embedded PCMs increases as the concentration of NPs in the suspension rises. Pumping power would increase only if PCMs were used to flow through pipes. However, in LTES systems, PCMs are held in spherical encapsulations and will not flow. This means that the viscosity augmentation will have no effect on the LTES system's performance.

6.3   PHASE CHANGE PROPERTIES OF NANOMATERIALS EMBEDDED PCMs

The latent heat and phase change temperatures of PCMs containing different mass fraction of TiO_2 NPs were determined through DSC analysis. The phase change temperatures and latent heats of nanomaterials embedded PCMs were also affected by the number of thermal cycles.

6.3.1   Thermal Properties of Nanomaterials Embedded PCMs

It has been discovered that the latent heats and phase change temperatures of stearic acid and nanomaterials with pure stearic acid and a variety of TiO_2 NPs have been

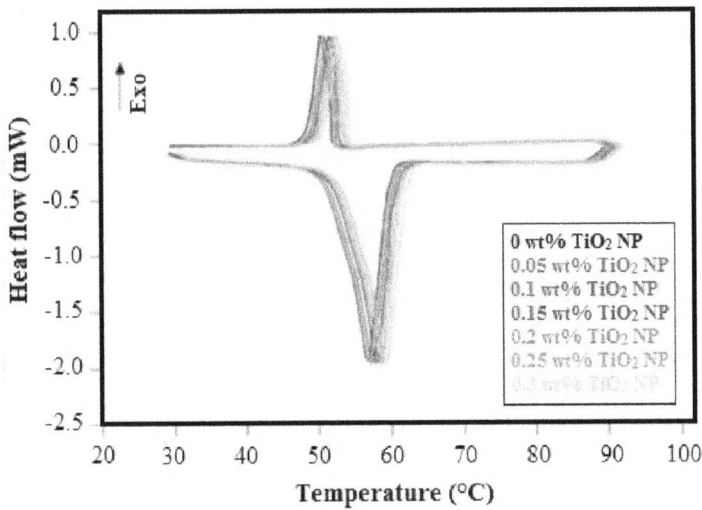

FIGURE 6.8 DSC measurements of stearic acid and nanomaterials embedded PCMs[1]

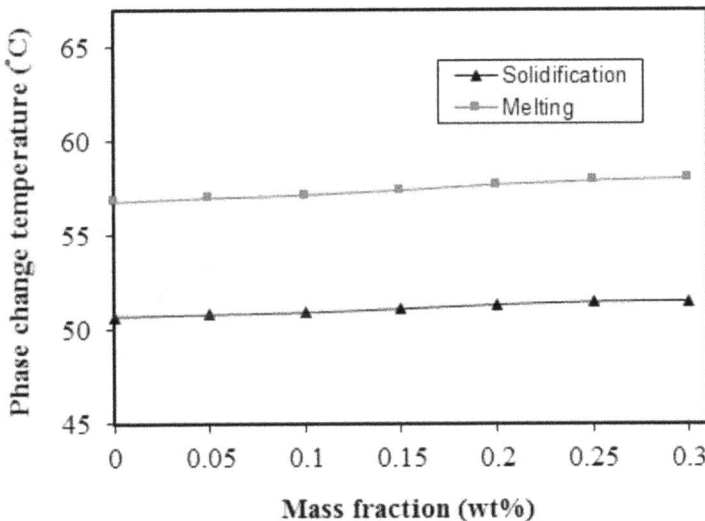

FIGURE 6.9 Effect of mass fraction of NPs on phase change temperature[1]

measured using DSC. Figure 6.8 shows the thermal characteristics of PCMs in relation to the mass percentage of NPs. The DSC results in Figures 6.9 and 6.10 reveal the phase change temperatures and latent heats for the samples[2].

The base material and the NPs contained PCMs have slightly different phase change temperatures. Latent heats of the base material and the NPs embedded PCMs

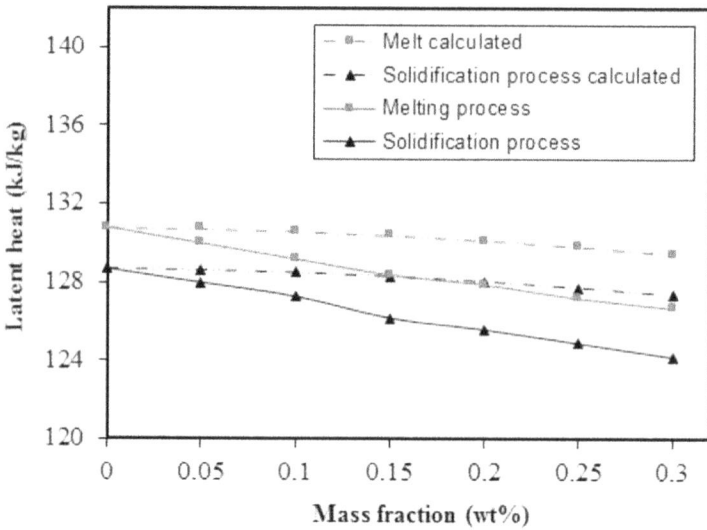

FIGURE 6.10   Effect of mass fraction of NPs on the latent heat[1]

both changed in similar ways. The scattered TiO_2 NPs and the base material have a strong physical contact, which causes these little modifications. The maximum variations in melting and solidification temperatures are −2.21% and −1.58%, respectively, as shown in Figure 6.9. Figure 6.10 shows the highest reductions in latent temperatures during melting and solidification[3].

The LTES system's performance is unaffected by these modest differences in the thermal properties of the NPs embedded PCMs. However, these changes may result in the energy storage and release rates of embedded NPs increasing compared to the existing material. Consequently, the minuscule decrease in latent temperatures of each PCM containing nanomaterials may be inconsequential. In addition, for solar energy storage, the PCMs that are commonly utilized have latent heat capacities of around 120 kJ/kg.

The latent heat of the nanomaterials in PCMs was estimated using the theory of mixture in comparison to the experimental data. Figure 6.10 clearly shows that the latent heat in the implanted PCM NPs is greater than the latent heat values discovered during DSC testing. This is owing to the NPs in the underlying material's surface shape, structure, and dispersion stability.

6.3.2   THERMAL RELIABILITY OF NANOMATERIALS EMBEDDED PCMS

PCMs will experience a maximum number of melting and solidification processes in long-term operations. The long-term thermal stability of the nanomaterials embedded PCM with 0.3wt% TiO_2 NPs was investigated. Figure 6.11 shows the variation of the nanomaterial phase change temperature with respect to the thermal cycle.

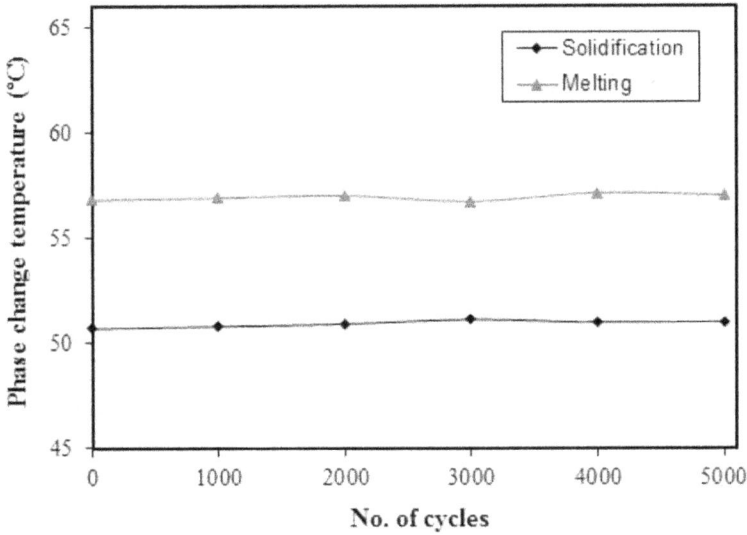

FIGURE 6.11 Variation of phase change temperature with respect to thermal cycle[1]

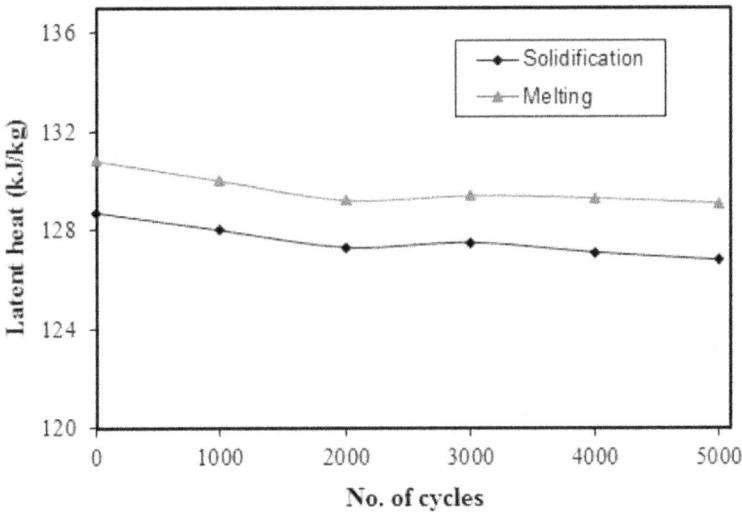

FIGURE 6.12 Variation of latent heat with respect to thermal cycle[1]

The maximum phase change temperature for solidification and melting were −0.59% and −0.35%, respectively. Figure 6.12 shows the latent heat variance for nanomaterials that are incorporated in PCM across thermal cycles. The latent temperatures for solidification and melting were found to vary by 1.47% and 1.29%, respectively.

The PCM's energy storage/release rates and the LTES system's performance will not be affected by these changes. Because of their latent heat storage capacity, these movements may be useful in solar heating applications. The average life cycle of about 13 years could be approximated from the standpoint of synthesized nanoparticles that were incorporated in PCMs. This resulted in a higher thermal stability for the TiO_2 NPs dispersed stearic acid as PCM.

6.4 PHASE CHANGE BEHAVIOR OF NANOMATERIALS EMBEDDED PCMs

Experiments were carried out to determine the increased thermal and heat transfer characteristics of the nanoparticles embedded PCMs during cooling and heating operations. PCM solidification and meting features in their pure state and with scattered TiO_2 NPs are also confirmed by the analytical solution and experimental data.

6.4.1 EXPERIMENTAL SETUP

Stearic acid and nanomaterials embedded PCMs containing 0.05, 0.1, 0.15, 0.2, 0.25, and 0.3wt% TiO_2 NPs were investigated individually in the TES tank using spherical encapsulation. The circulating HTF was water with a temperature range of 20°C to 90°C.

Prior to conducting the experiment, the temperature controllers in each of the tanks were maintained at the same level. The temperature changes through the melting and solidification process were also measured using thermocouples attached to the PCM encapsulation. The digital camera was used to observe the solidification and melting processes of the nanoparticles containing PCMs. To provide better operation features of the LTES system, with varying thermal loads, 80 channel data loggers (Agilent 34972A, USA) were attached to all thermocouples. Thermocouples were used to monitor the conditions of the tanks, which were being tested, using a data logger.

To cool the circulating HTF to the appropriate temperature for studying the solidification process of the PCM and nanomaterial embedded PCMs, the refrigeration system was employed. The HTF temperature was kept at 30°C during the chilling process and pumped to the TES tank with the circulating pump. Cold energy was transferred from the HTF to the PCMs until they became solidify completed. Similarly, the heating system was used to heat the HTF from 30°C to the appropriate temperature for studying the melting process of the PCM and nanomaterials embedded PCMs. During the melting process, the HTF temperature was fixed at 85°C and the circulating pump was used to supply the HTF to the TES tank. The heat energy from the HTF was subsequently transferred to the PCMs, causing them to melt completely. To assure measurement repeatability, the melting and solidification operations for the PCM and nanomaterials embedded PCMs were done five times in this manner[1, 3].

6.4.2 SOLIDIFICATION AND MELTING CHARACTERISTICS

The heat transfer characteristics of stearic acid and the PCMs with incorporated NPs were studied during the melting and solidification processes. The stearic acid and PCMs with integrated nanomaterials have corresponding temperature curves in

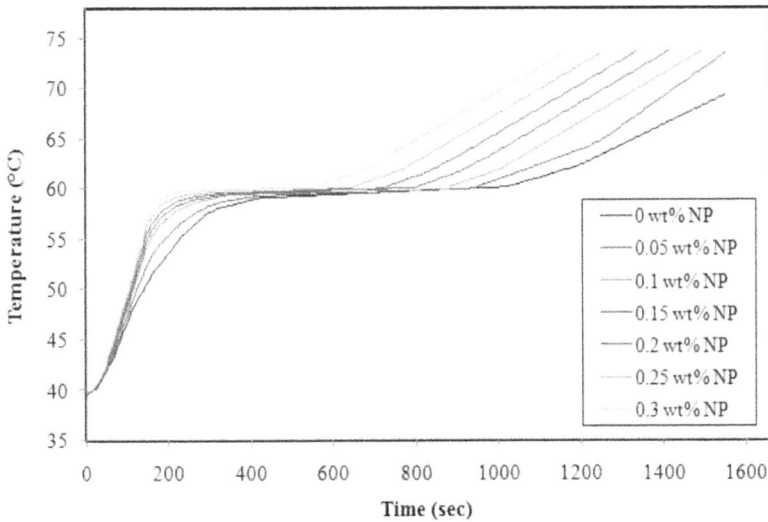

FIGURE 6.13 Melting processes of nanomaterials embedded PCMs[1]

Figure 6.13. Stearic acid and NPs embedded PCMs started at 39°C during the melting process, as depicted in Figure 6.13. Stearic acid and nanomaterials embedded PCMs' temperatures increase with time, and the increase continues until the stearic acid and nanomaterials embedded PCMs melt.

As the melting point was reached, the PCMs, containing stearic acid and NPs, were observed to transition from a solid to a liquid phase. Complete melting times for stearic acid and NPs embedded PCM with 0.3wt% TiO_2 are 995s and 560s, respectively. When compared to pure stearic acid, time savings of nanomaterials embedded PCMs with 0.05, 0.1, 0.15, 0.2, 0.25, and 0.3wt% for complete melting were determined to be 7.03, 12.56, 19.59, 28.64, 35.17, and 43.72%, respectively. During the melting process, it was clear that the dispersed NPs, which were delivered as stearic acid, enhanced conductive heat transfer as well as convective heat transfer.

Figure 6.14 shows that stearic acid and PCMs with embedded NPs started at 67°C. Stearic acid and NPs contained in PCMs cool over time, until they reach their freezing point. The PCMs embedded with stearic acid and NPs are quickly going from liquid to solid as soon as they approach freezing point. Stearic acid and NPs embedded PCMs with 0.3wt% TiO_2 NPs take 1510s and 885s, respectively, for complete solidification from the start.

When compared to pure stearic acid, a composite PCM that completely solidifies with 0.05, 0.1, 0.15, 0.2, 0.25, and 0.3wt% for a respective reduction in time is 6.62, 13.57, 20.53, 26.82, 34.11, and 41.39%, respectively. These findings are in agreement with computational simulation studies on the increase in thermal conductivity of NPs that are placed in PCMs for thermal energy storage systems. The solidification of stearic acid with dispersed effective NPs enhanced the conductive heat transfer

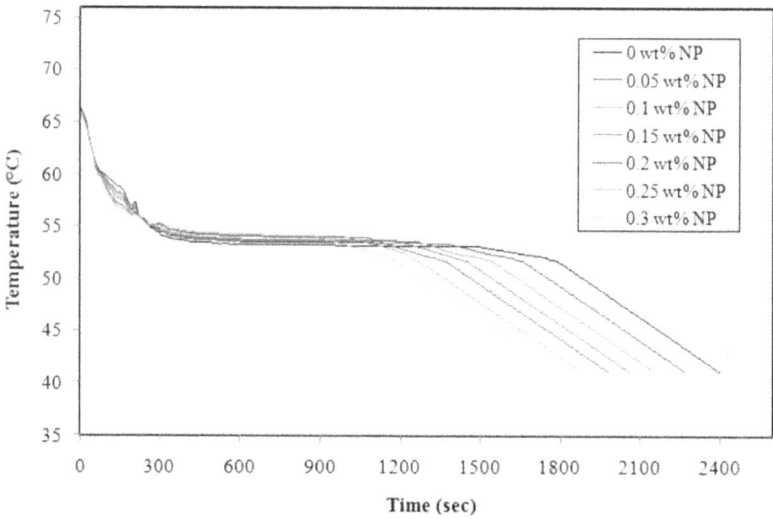

FIGURE 6.14   Solidification processes of nanomaterials embedded PCMs[1]

even further. According to the results of the experiment, the melting and solidification rates of nanomaterials embedded in PCMs rise as the mass fraction of TiO_2 NPs in stearic acid increases. The melting and solidification rates of the NPs that are encased in PCMs are noticeably improved by the thermal conductivity of stearic acid.

Figures 6.15 and 6.16 represent the melting and solidification processes of stearic acid with 0.3 wt% of TiO_2 NPs as nanomaterials. As shown in Figure 6.15, when HTF is applied to the outside of the beaker, nanomaterials held in PCM tend to melt near the inner wall of the beaker. Nanomaterials embedded PCM melts further as heat is applied to the beaker's outer wall on a continual basis, and liquid PCM contains the solid PCM in a ring shape, as illustrated in Figure 6.15. The diameter of the ring form solid PCM reduces fast as the melting process progresses, until the ring form solid PCM completely disappears. It is clear at this moment that the nanomaterials incorporated in PCM have reached the complete melting stage.

FIGURE 6.15   Photographic views for complete melting progress of composite PCM[1]

FIGURE 6.16 Photographic views for complete solidification progress of composite PCM[1]

The PCM in the beaker's inner wall solidifies during the solidification process of the embedded NPs, as is seen in Figure 6.16. As the PCM solidifies, the area of the solidified NPs inside the beaker expands and moves closer to the beaker's center. As is shown in Figure 6.16, the hardened PCM in the beaker's middle after the process of melting had ended. Furthermore, during melting, heat is transferred from HTF to solid PCM, but during solidification, heat is transferred from molten PCM to HTF.

6.4.3 SOLIDIFICATION AND MELTING ANALYSES

Figure 6.17 depicts the time-dependent mobility of the PCM's solid-liquid interface and incorporated nanomaterials during solidification and melting. Figure 6.17a shows the analytical solution and experimental findings for the chilling cycle, which reveal that the nanomaterials embedded PCMs solidification time is significantly shorter than the base material.

The reduction in solidification time is due to two main phenomena: (1) thermophoretic diffusion and (2) heterogeneous nucleation. At an arbitrary radial distance, many well dispersed NPs are expected to exist in the nanomaterials embedded PCMs. The thermally conductive NPs present near the wall surface absorb the cold energy efficiently and transfer it to the adjacent NPs embedded PCMs in this circumstance. Meanwhile, the NPs in nanomaterials embedded PCMs stimulate the heterogeneous nucleation process, which allows the establishment and subsequent expansion of the stable nucleus.

The NPs provide the activation energy needed to keep the ice crystals from growing out of control. As a result, the solid-liquid interface propagates faster, resulting in immediate freezing at the expense of exothermic heat release from nanomaterials contained in PCMs. Similarly, the findings of the heating cycle shown in Figure 6.17b suggest that the melting of the nanomaterials embedded PCMs is aided mostly by the NPs' nanoconvection. The higher the concentration of NPs, the higher the rate of phonon-like heat transfer that occurs for rapid freezing and melting of nanomaterials embedded PCMs. With higher concentrations of TiO_2 NPs, the solidification and melting durations of nanomaterials embedded PCMs were dramatically lowered, according to the analytical solution and experimental results.

As the mass fractions of TiO_2 NPs were increased from 0.1% to 0.3%, the times needed for complete solidification and melting were dropped from 1710 s to 1270 s

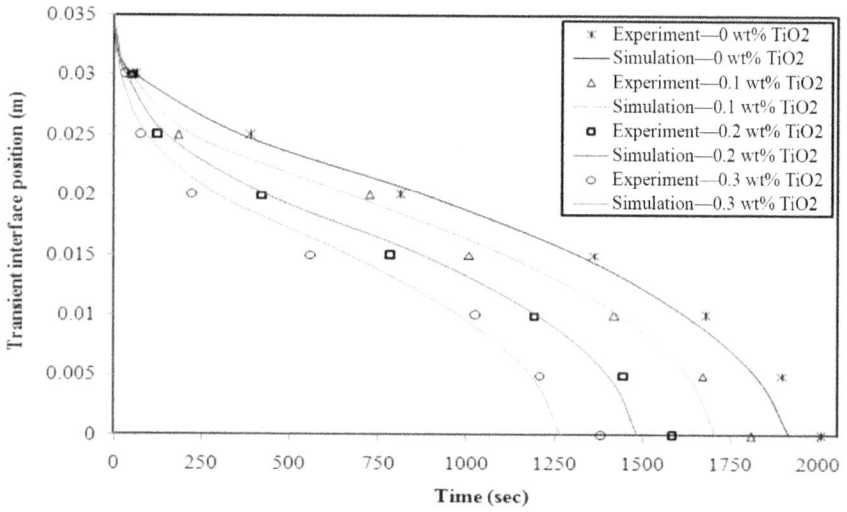

(a)

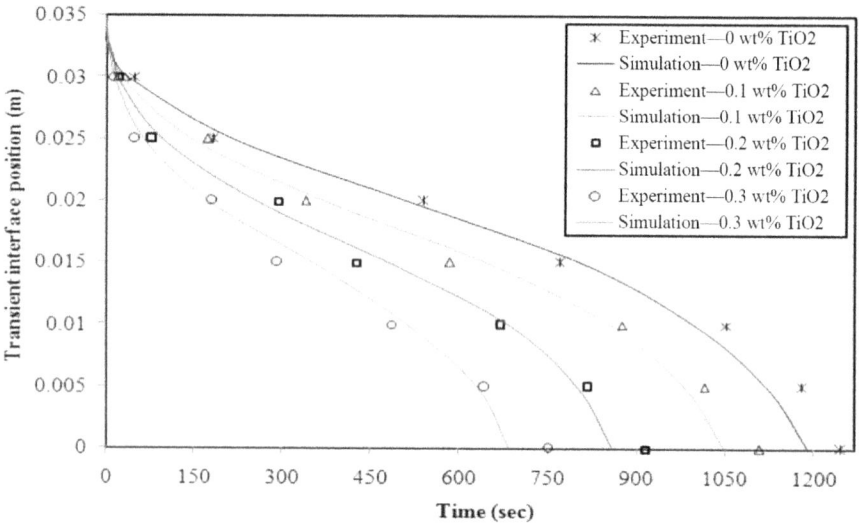

(b)

FIGURE 6.17   Comparison between experimental and analytical results for (a) solidification
(b) melting[1]

and 1048 s to 684 s, respectively. Experiments also show that the time it takes for nanomaterials embedded PCMs to completely freeze and melt was reduced from 1810 s to 1385 s and 1110 s to 750 s, respectively. For the complete solidification and melting processes, the analytical solution and experimental data had an average deviation of 8.31% and 8.84%, respectively.

6.5 PALMITIC ACID-TiO$_2$ NANOFLUID PCM

The synthesis and characterization of palmitic acid-TiO$_2$ nanofluid PCM are described in this section. The solidification and melting characteristics of the integrated PCMs were also examined for the purpose of evaluating the enhanced thermal and heat transfer performance of the LTES systems. The transient interface positions of nanomaterials embedded in the PCM are investigated theoretically and empirically during cooling and heating activities.

6.5.1 CHARACTERIZATION OF NPs

Figure 6.18 shows a TEM picture of TiO2 NPs, with diameters ranging from 17 nm to 69 nm for the as synthesized particles. Some NPs were discovered to be aggregated in the TEM picture, which could be decreased using the ultrasonic vibrator and the addition of surfactant. The spherical shape of the NPs is confirmed by the SEM picture in Figure 6.19, and this spherical shape could provide appropriate contact surface for heat flow.

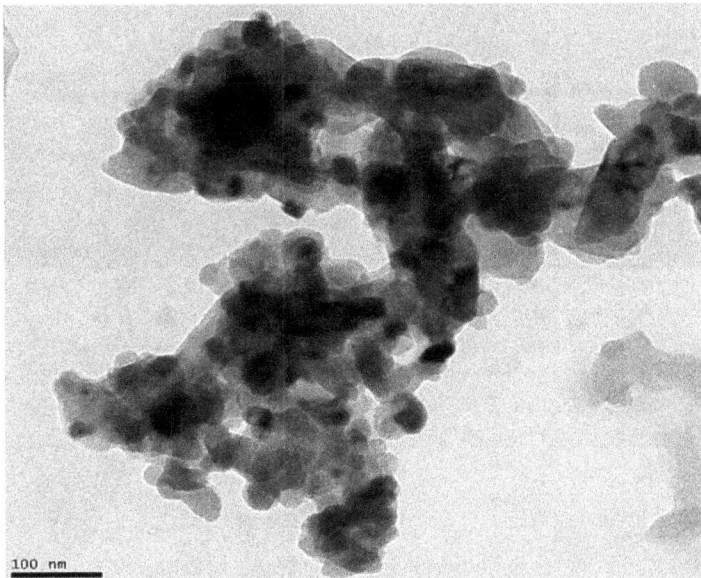

FIGURE 6.18 TEM image of TiO$_2$ NPs[1]

200 nm   EHT = 20.00 kV   Signal A = InLens
          WD = 4.2 mm   Mag = 50.37 K X

FIGURE 6.19   SEM image of TiO$_2$ NPs[1]

6.5.2   PREPARATION OF NANOMATERIALS EMBEDDED PCMS

Thermofisher Laboratories Scientific Private Limited, India, provided palmitic acid (melting point 60–62°C). Table 7.3 shows the thermophysical characteristics of palmitic acid. The nanomaterials embedded PCMs of 0.1, 0.2, and 0.3wt%. TiO$_2$ NPs were prepared using a two-step technique. During the preparation of nanomaterials embedded PCMs, it was vital to include certain properties such as continuous dispersion, low particle agglomeration and no chemical fluid modification. Sodium dodecylbenzene sulfonate (SDBS) was the capping agent utilized for the uniform dispersion of NP in palmitic acid.

The technique of sonication has been utilized to improve the stability of particle distribution and reduce agglomeration with an ultrasonic frequency at 40 kHz. Residing times of nanofluid PCMs were 35, 40, and 45 min, respectively, for mass fractions of 0.1, 0.2, and 0.3wt%.

6.5.3   INFLUENCE OF NPS ON THE THERMAL CONDUCTIVITY

Palmitic acid's thermal conductivity with diverse mass fractions of TiO$_2$ NPs is anticipated to be higher than pure palmitic acid. The thermal conductivity of nanofluid

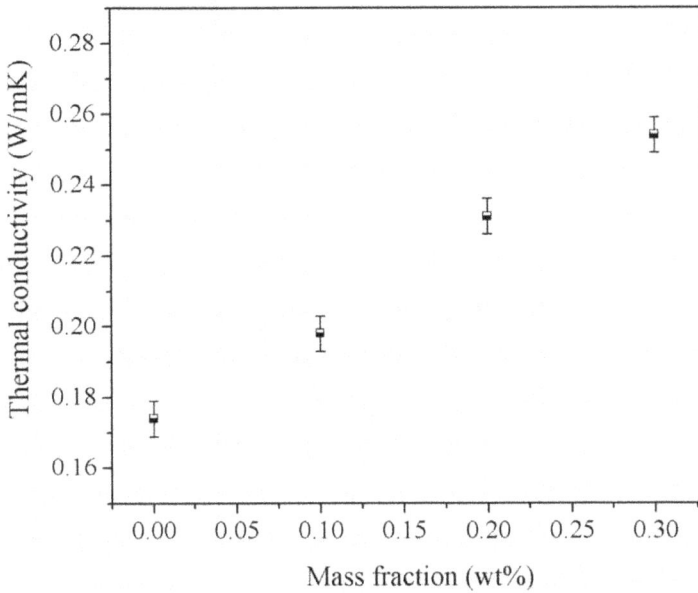

FIGURE 6.20 Influence on the thermal conductivity of the mass fraction of nanoparticles[1].

PCMs depending on the mass fraction of NPs is shown in Figure 6.20. The thermal conductivity of nanofluid PCMs increased by 0.198, 0.231, and 0.256 W/mK, respectively, when 0.1, 0.2, and 0.3wt% TiO_2 NPs were added.

The thermal conductivity of the base fluid also increases when the TiO_2 concentration in the fluid increases. Furthermore, the nanofluid PCMs' thermal conductivity was shown to be linear. Thermal conductivity enhancement would be nonlinear as the mass fraction of the NPs increased.

The optimization of the mass fraction of NPs is essential to achieve maximum thermal conductivity of nanofluid PCMs. In comparison to pure palmitic acid, the thermal conductivity of TiO_2-palmitic acid nanofluids increased by 13.79, 32.18, and 46.55% for 0.1, 0.2, and 0.3wt% NPs, respectively. Due to increasing heat conductivity, nanofluid PCMs melts and solidifies more quickly than palmitic acid. Palmitic acid-TiO_2 NPs also worked as a nucleating agent, minimizing the super-cooling effect and serving to melting and solidification in the PCM.

6.5.4 PHASE CHANGE PROPERTIES OF NANOMATERIALS EMBEDDED PCMs

The phase change temperatures and latent heating of implanted NPs for solidification and melting was measured using DSC analysis with various mass fractions of TiO_2 NPs. Similarly, the number of thermal cycles had an effect on the phase change temperatures and latent heats of nanomaterials embedded PCMs.

6.5.5 THERMAL PROPERTIES OF NANOMATERIALS EMBEDDED PCMs

Phase change temperature was determined using DSC measurements during the melting and solidification of 0.1, 0.2, and 0.3wt% NPs of base material and PCM nanomaterials. Figures 6.21 and 6.22 show phase change temperatures and latent

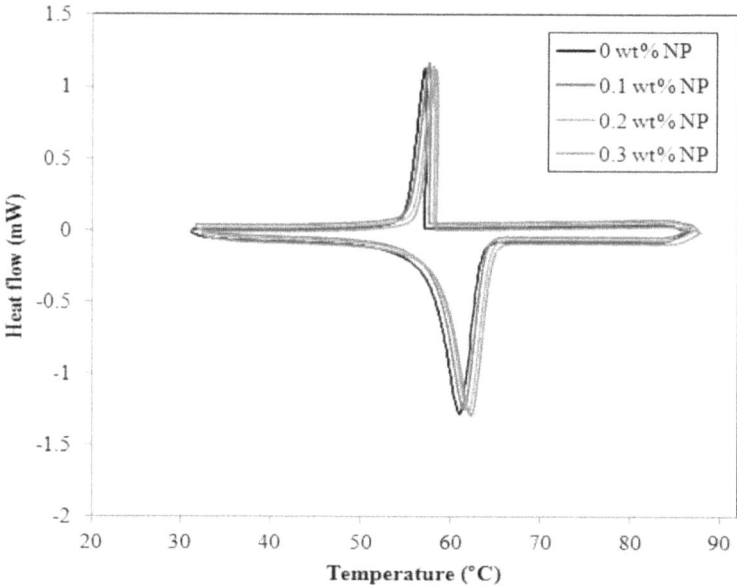

FIGURE 6.21   DSC measurements of nanomaterials embedded PCMs[1].

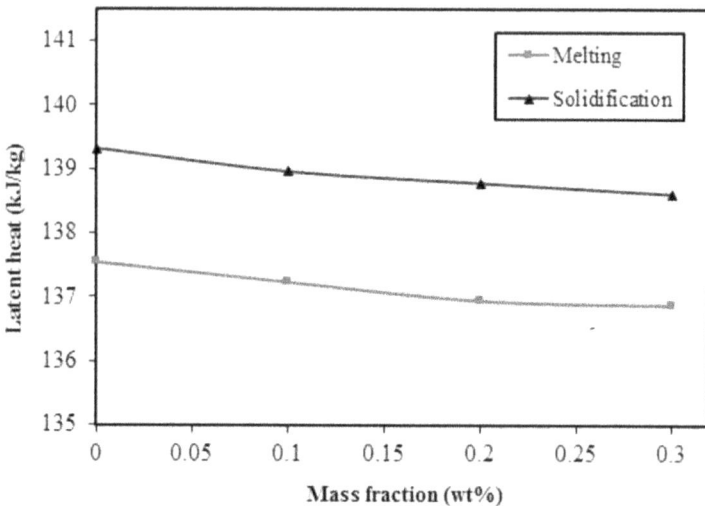

FIGURE 6.22   Effect of concentration of NPs on latent heat of PCM[1].

heats calculated using DSC readings. Small differences in phase change temperatures of the base material and NPs embedded PCMs were discovered. The latent heat of the base material and the NPs embedded PCMs showed similar variations. The strong physical contact between TiO_2 NPs and the base material is responsible for these minor modifications[4].

Figure 6.22 shows the maximum drop in latent heat during melting and solidification. Despite these little variations in the thermal properties of the embedded nanomaterials PCMs, the LTES system's performance would never be significantly harmed. These modifications, on the other hand, improved the energy storage and release processes of the NPs-incorporated PCMs over the base material. For heating applications, solar energy storage PCMs with a latent heat capacity of around 120 kJ/kg are usually preferred. Due to their low heat output, nanomaterials' latent heat loss could be ignored.

6.5.6 THERMAL RELIABILITY OF COMPOSITE PCMS

PCMs should maintain the same phase change temperatures and latent heats over many thermal cycles for best long-term service. As a result of these modifications, the LTES system's ability to store and release energy would be negatively affected, resulting in a decrease in performance. For verifying the long-term thermal stability, the nanomaterials embedded PCM containing 0.3wt% TiO_2 NPs was examined.

In Figure 6.23, the range of phase change temperature for nanomaterials is shown with a view to the number of heat cycles. The maximum increases in solidification and melting temperatures were seen at −1.34% and −1.59%, respectively. Figure 6.24 shows the change in latent heat during composite PCM thermal cycles. The highest

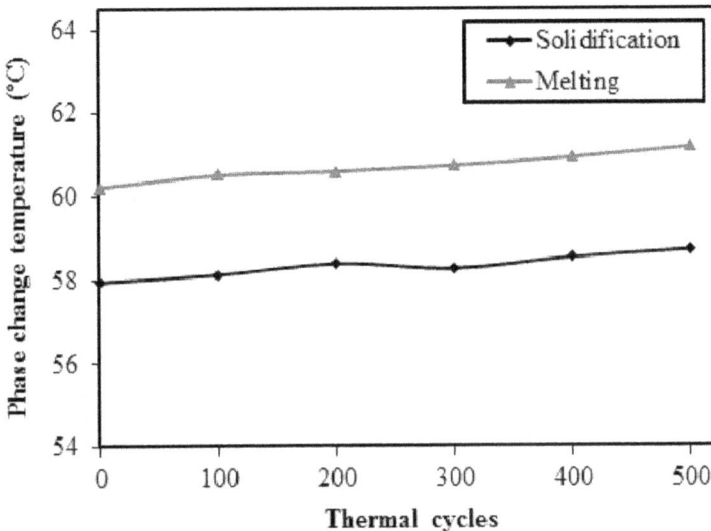

FIGURE 6.23   Variation of phase change temperatures with respect to thermal cycle[1].

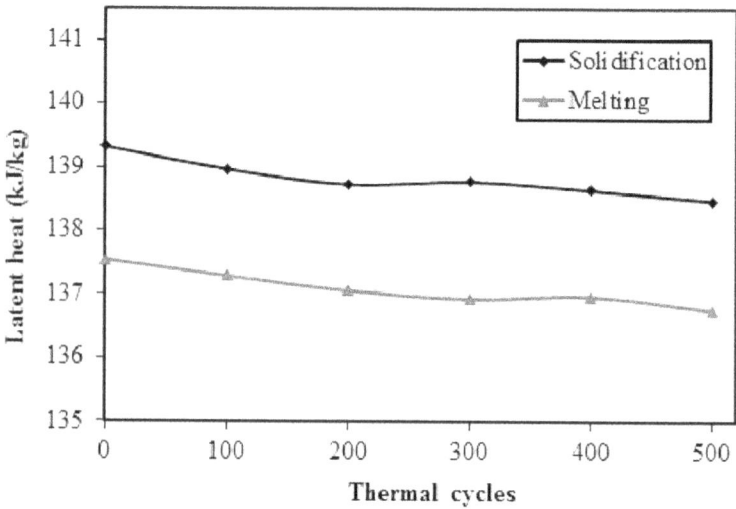

FIGURE 6.24   Variation of latent heats with respect to thermal cycle[1].

recorded temperature change occurred in the solidification and melting of solids, at 0.61% and 0.56%, respectively.

Thermal energy storage and release rate modifications would not have a negative impact on the LTES system's performance and would not impede the system's ability to function. Therefore, these modifications are acceptable in order to achieve solar heating. Thus, the thermal dependability of palmitic acid based TiO_2 nanofluid PCM is superior to that of the base material.

6.6   PHASE CHANGE BEHAVIOR OF NANOMATERIALS EMBEDDED PCMs

Experimental studies have been used to determine the improved thermal and heat transfer performance of NPs integrated into refrigeration and heating processes. Experimental results and an analytical solution relating to the solidification and mixing features of the PCM in its pure form and with distributed TiO_2 NPs are also confirmed.

6.6.1   Experimental Setup

A spherical encapsulation with the assistance of TiO_2 NPs (0.1, 0.2, 0.3 wt%) was used to test PCMs embedded with palmitic acid. Water is used as a circulation heat transfer fluid whose temperature can be varied from 30°C to 90°C. The cooling system was used for the base material and the nanomaterials in the PCMs to attain the temperature level necessary in order to test the solidification process[1].

The HTF was kept at 46°C and pumped to the TES tank with the circulating pump during the cooling process. HTF had the cold energy when the liquid was transferred to the PCMs until it was entirely solidified. The heating system has been switched on to improve the HTF circulating temperature in the right condition to examine the melting process in the basic material and embedded NPs of PCMs. The temperature of the TES tank was 74°C and the HTF was circulating through the circulating pump. The heat energy from HTF was transferred to the PCMs until the whole melting point was reached. The melting and solidifications for the base material and embedded PCM were done five times to confirm the correctness of the measured results.

6.6.2 SOLIDIFICATION AND MELTING CHARACTERISTICS

The heat transfer rates of palmitic acid and nanomaterial embedded PCMs during melting and solidification were assessed. Figure 6.25 shows the temperature profiles of the PCMs for the melting process, measured at the center of the spherical encapsulation. Palmitic acid and embedded nanomaterials in PCMs reached a temperature of 40°C at the start of the melting process, as depicted in Figure 6.25. The base material and nanomaterials embedded in PCMs increase with time and last until the base material and nanomaterials embedded PCMs reach a melting point.

After base material and nanomaterial PCMs have reached the melting point, they start to convert from solid to liquid. The time needed to fully smelter the base material from the starting point is 855, whereas nanomaterial PCMs at 0.1, 0.2, and 0.3wt are 800s, 730s and 675s, respectively. It is evident that the conducting and convective heat transfers are both created in the base material due to the scattered NPs[4].

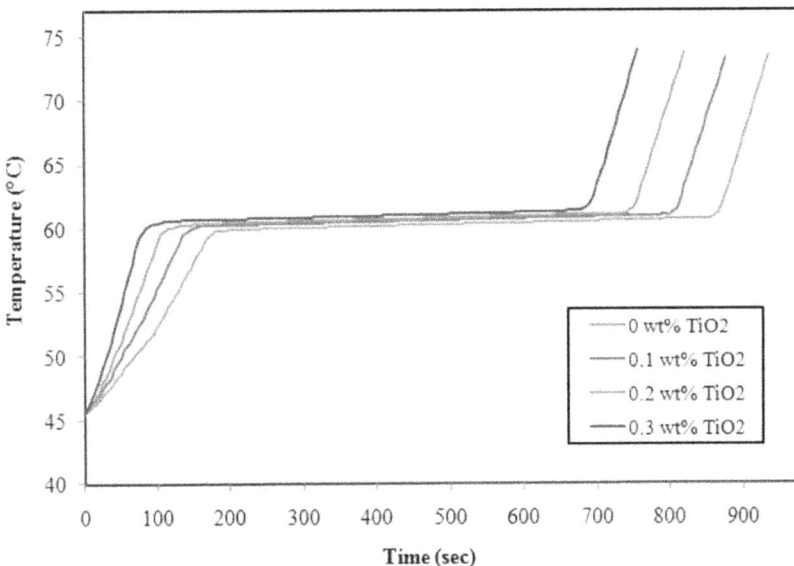

FIGURE 6.25 Melting curves of nanomaterials embedded PCMs[1].

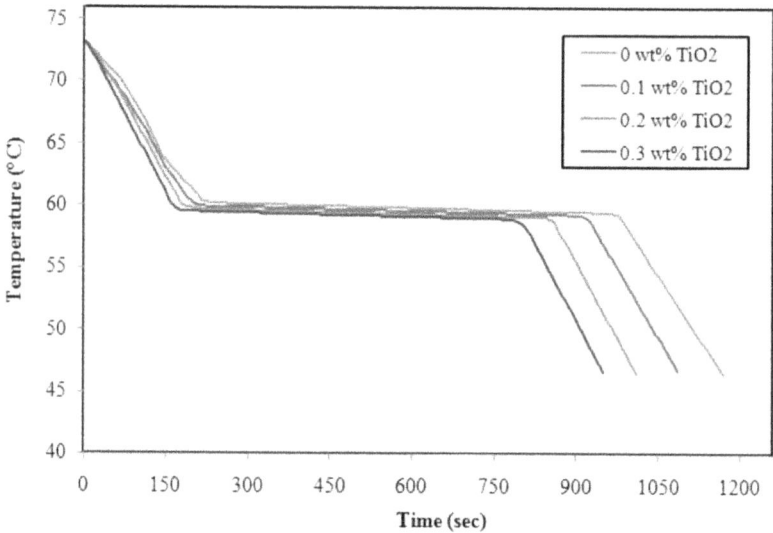

FIGURE 6.26 Solidification curves of nanomaterials embedded PCMs[1].

At the beginning of the solidification process temperatures for the base material and nanomaterials were 80°C (as shown in Figure 6.26). The base material and NPs steadily incorporate PCMs as time continues, with the process continuing until they strike their freezing point. After reaching the freezing point, the base material and NPs incorporated in PCMs have shifted to a solid phase. The base material's entire solidification time from the starting point is 970 s, while the nanomaterials in PCMs at 0.1, 0.2, and 0.3wt% have 910s, 850s, and 775s solidification time.

With these findings, the thermal conductivity of nanofluid PCMs for thermal energy storage systems was found to have a significant increase. The overall heat transfer increased because of the dispersed effective NPs introduced into the base material during the solidification process. The experimental results show that the amount of TiO_2 NPs in the base material increases, as do the melting and solidification speeds of the nanomaterials mixed into the PCMs. In addition, the improvement in heat conductivity is obvious since NPs embedded PCMs melt and solidify faster.

6.6.3 Solidification and Melting Analyses

Figure 6.27 presents the time-dependent mobility of the solid-liquid interface in solidification and melting of PCMs and nanomaterial embedded PCMs. Figure 6.27a indicates that the incorporated nanomaterials in the PCMs decrease the time to solidify, but the base material takes longer.

Two key factors were responsible for the shortening of the solidification time: (i) thermophoretic diffusion and (ii) heterogeneous nucleation. The embedded PCMs' nanomaterials contain several well-dispersed NPs at adjustable radial distances. The

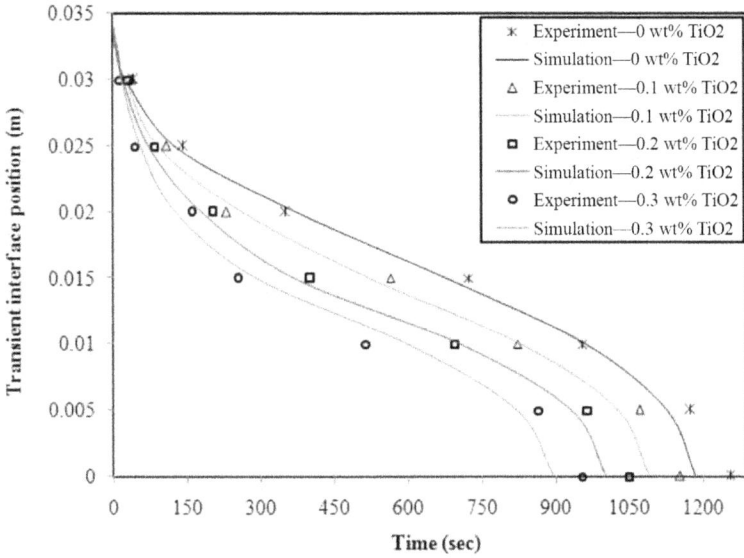

(a)

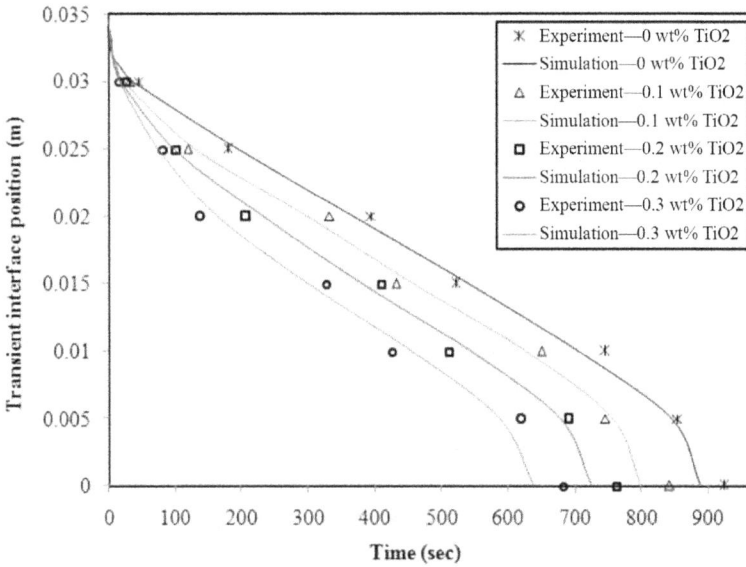

(b)

FIGURE 6.27   Comparison between analytical and experimental results for (a) solidification, (b) melting[1, 5].

thermally conductive NPs are therefore able to efficiently absorb the cold energy that is present at the surface and transfer it to the neighboring NPs embedded PCMs. Meanwhile, the nanomaterials in embedded PCMs cause heterogeneous nucleation. This allows the development and growth of a stable nucleus[6].

As a result, freezing occurs faster, but there is a resulting exothermic heat release of nanomaterials within PCMs. Figure 6.27b illustrates that melting of the nanomaterials in the PCMs was increased by the nanoconvection of the NPs. The ability to rapidly freeze and melt PCMs with embedded nanomaterials depends on the quantity of NPs in the system. Maximum numbers of NPs provide higher rates of phonon-like heat transfer. The computational results and experimental findings confirm that the nanomaterial-based PCMs solidified and melted more quickly with higher concentrations of TiO_2 NPs.

The analytical result indicated that the times for complete solidification and melting decreased from 1090 s to 894 s and from 800 s to 638 s for NPs with mass fractions of 0.1% and 0.3%, respectively. The freezing and melting times of the nanomaterial-containing PCMs were found to be reduced from 1152 s to 952 s and 842 s to 682 s. An overall average deviation of 6.09% and 6.45% were noted for the whole of the solidification and melting processes, respectively[5, 7, 8].

6.7 SUMMARY

By dispersing various mass percentages of TiO_2 NPs in stearic acid and palmitic acid independently, an innovative nanomaterial embedded PCM was developed to investigate improved thermal conductivity for solar energy storage. For maintaining a consistent dispersion of TiO_2 NPs in PCMs, SDBS was used as the surfactant. TEM imaging and XRD analysis found that the TiO_2 NPs, produced by the sol-gel technique, had an average size that agreed with the results. An FTIR investigation found that the stearic acid-TiO_2 nanofluid PCM was characterized by its excellent chemical compatibility and strong physical contact.

For solidification and melting, the nanomaterial embedded PCMs' phase change temperatures and latent heats were evaluated by the use of DSC analysis. It was shown that small variations existed between nanomaterials within the basic materials and PCMs after DSC analysis. The LTES system's heat transfer performance did not suffer from these slight variations. Thermal conductivities of nanomaterial embedded PCMs were higher than PCMs without nanomaterials in both studies. No notable changes in the LTES systems' performance occurred due to NPs' viscosity enhancements.

According to the results of the experiments, it was found that stearic acid-TiO_2 nanofluid PCMs had a time savings of 7.04, 12.56, 19.59, 28.64, 35.17, 43.71%, and 6.62, 13.58, 20.53, 26.82, 34.11, 41.39% for complete melting and solidification processes when 0.05, 0.1, 0.15, 0.2, 0.25, 0.3wt% TiO_2 NPs were used. In addition, the time needed to completely melt and solidify PCMs with 0.1, 0.2, 0.3 wt% NPs was 6.43, 14.62, 21.05%, and 6.18, 12.37, 20.11% correspondingly. In addition, the validity of the analytical solutions for solidification and melting of the basic PCMs and nanomaterials-incorporated PCMs was verified by experimental data.

REFERENCES

[1] Harikrishnan, S., Deenadhayalan, M. & Kalaiselvam, S., 'Experimental investigation of solidification and melting characteristics of composite PCMs for building heating application', *Energy Conversion and Management*, vol. 86, pp. 864–872, 2014.

[2] He, Q., Wang, S., Tang, M. & Liu, Y., 'Experimental study on thermophysical properties of nanofluids as phase-change material (PCM) in low temperature cool storage', *Energy Conversion and Management*, vol. 64, pp. 199–205, 2012.

[3] Kenisarin, M. & Mahkamov, K., 'Solar energy storage using phase change materials', *Renewable and Sustainable Energy Reviews*, vol. 11, pp. 1913–1965, 2007.

[4] Maghrabie, H. M., Elsaid, K., Sayed, E. T., Radwan, A., Abo-Khalil, A. G., Rezk, H., . . . Olabi, A. G., 'Phase change materials based on nanoparticles for enhancing the performance of solar photovoltaic panels: A review', *Journal of Energy Storage*, vol. 48, p. 103937, 2022.

[5] Younes, H., Mao, M., Murshed, S. S., Lou, D., Hong, H. & Peterson, G. P., 'Nanofluids: Key parameters to enhance thermal conductivity and its applications', *Applied Thermal Engineering*, p. 118202, 2022.

[6] Na, Y. S., Kihm, K. D. & Lee, J. S., 'Opposite ReD-dependencies of nanofluid (Al2O3) thermal conductivities between heating and cooling modes', *Applied Physics Letters*, vol. 101, no. 8, p. 083111, 2012.

[7] Wu, S., Zhu, D., Zhang, X. & Huang, J., 'Preparation and melting/freezing characteristics of Cu/Paraffin nanofluids as phase-change material (PCM)', *Energy and Fuels*, vol. 24, pp. 1894–1898, 2010.

[8] Harikrishnan, S. & Kalaiselvam, S., 'Preparation and thermal characteristics of nanomaterials embedded phase change materials for thermal energy storage', Shodh ganga, 2017. http://hdl.handle.net/10603/142206.

7 Composite PCMs for Thermal Energy Storage System

CONTENTS

7.1 INTRODUCTION

Despite extensive research into the improved thermal conductivity of PCMs through the use of only one type of NPs with varying mass or volume fractions, few studies

DOI: 10.1201/9781003163633-7

have examined the enhanced thermal properties of PCMs that are dispersed with hybrid NPs and other NPs. Such studies are certainly useful in deciding which NPs should be scattered in PCMs, allowing thermal conductivity to be improved even further. In this chapter, the ability to store thermal energy in a thermal energy storage device was examined in two different PCMs.

The base material for the first investigation was paraffin, and the supporting material was the $(CuO-TiO_2)$ hybrid nanomaterial used to construct the composite PCM. According to the results of another investigation, it was shown to be effective to use a lauric acid (LA) and stearic acid (SA) based base material with dispersed TiO_2, ZnO, and CuO NPs in a mass fraction of 1 wt% of the base material. Surfactants and capping agents used to stabilize NPs in the base materials were SDBS in both composite PCMs.

The number of thermal cycles and mass fraction were used to estimate latent heat and phase change temperatures in the composite PCMs. The augmentation of the composite PCMs' thermal conductivity and viscosity owing to the addition of NPs was measured[1, 2].

7.2 COMPOSITE PCM USING HYBRID NANOMATERIALS

To measure the thermal conductivity enhancement, paraffin-based hybrid nanomaterials with 0.25, 0.5, 0.75, and 1.0 wt% mass fractions were distributed. The thermal conductivity of PCMs containing hybrid nanomaterials was compared using CuO NPs and TiO_2 NPs with mass fractions of 0.25, 0.5, 0.75, and 1.0wt% in paraffin.

7.2.1 CHARACTERIZATION OF HYBRID NANOMATERIALS

Hybrid nanomaterials were dispersed in the base fluid in a range of sizes, as shown in Figure 7.1. In the particle size analyzer, a single peak was identified, indicating that the hybrid nanomaterials in the base fluid were dispersed equally throughout the fluid.

Figure 7.2 shows that hybrid nanomaterials have an average size of 21 nm and can range in size from 11 nm to 42 nm. Figures 7.2a–d show SEM images of CuO, TiO_2,

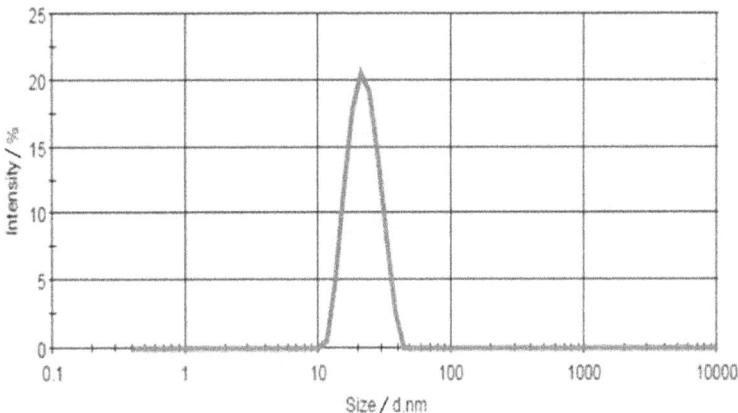

FIGURE 7.1 Particle size distribution of hybrid nanomaterials in the base fluid[3].

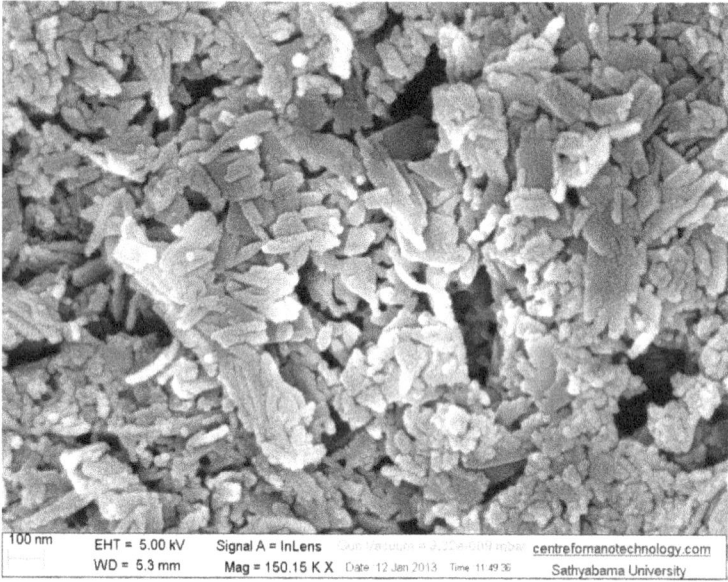

(a)

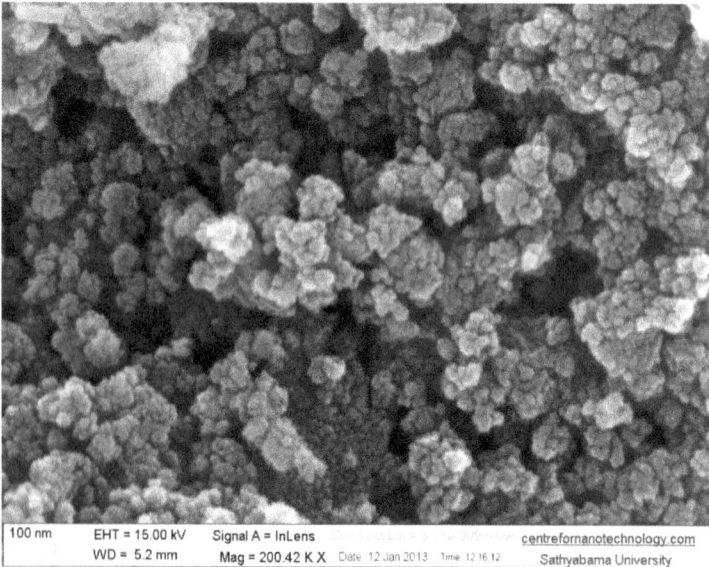

(b)

FIGURE 7.2   SEM images (a) CuO NPs, (b) TiO$_2$ NPs, (c) CuO-TiO$_2$ hybrid nanomaterials, (d) paraffin with 1.0wt% hybrid nanomaterials[3].

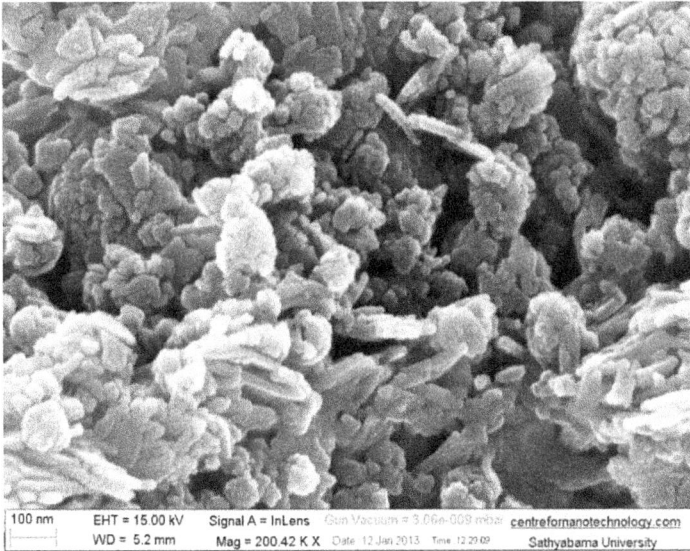

(c)

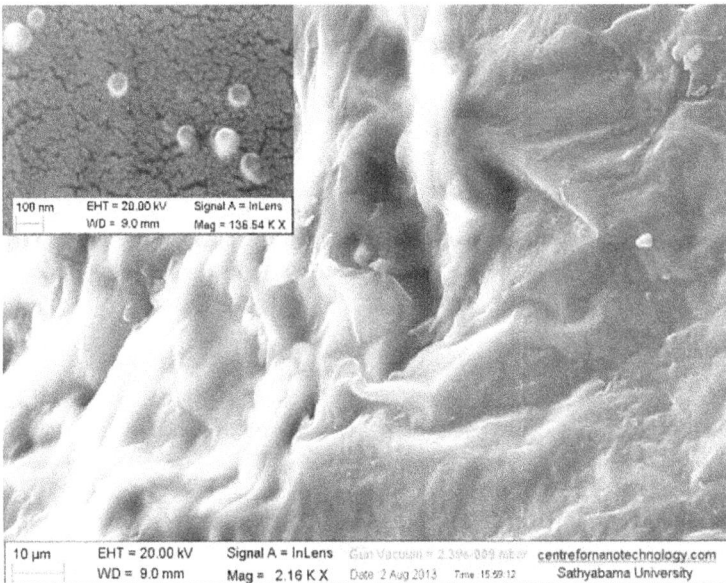

(d)

FIGURE 7.2 (Continued)

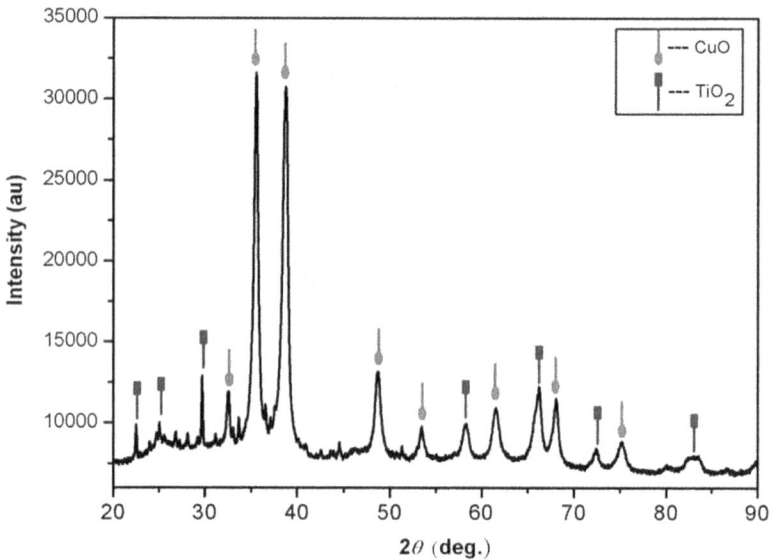

FIGURE 7.3   XRD pattern for hybrid nanomaterials[3].

hybrid nanomaterials, and paraffin containing 1.0% hybrid nanoparticles. In both the CuO and TiO_2 NPs, the rod and spherical shapes were found to be consistent. As a result of their rod-like structure, CuO NPs have a greater heat-contact surface area than TiO_2's spherical shape. The rod-shaped CuO NPs dispersed in the base material, on the other hand, will increase the base material's viscosity. Since it is crucial to use an optimal amount of rod-shaped NPs in the base material, it is important to spread them evenly. The spherical TiO_2 nanostructures appear to be near and on the CuO rod nanostructures as seen in Figure 7.2c. It was discovered that the distribution of the hybrid nanomaterials in the paraffin was robust diffused.

Figure 7.3 shows the peaks of TiO_2 and CuO NPs, represented by different colors. The XRD examination shows that there are no pseudo peaks in the pattern. The XRD pattern of hybrid nanomaterials also reveals the presence of CuO and TiO_2 NPs. In addition, it had good concordance with JCPDS files for CuO and TiO_2 NPs, with 45–0937 and 74–1940, respectively.

7.2.2   PREPARATION OF COMPOSITE PCMS

It is possible to make PCMs with varied mass fractions, such as 0.25, 0.5, 0.75, and 1.0wt% of hybrid nanomaterials as a base material and a support material. Paraffin was acquired from RFCL in India and has a melting point of 60 to 62°C. Table 7.1 details the thermophysical characteristics of the paraffin. For the base material, the CuO and TiO_2 NPs were dispersed by a 1:1 mass ratio (i.e. 50% CuO/50% TiO_2) in 0.25, 0.5, 0.75, and 1.0wt%. Nanomaterials were suspended and dispersed more effectively in the base material by using SDBS as a surfactant and it was sourced

TABLE 7.1
Thermophysical Properties of the Paraffin[3]

S.No	Properties of the PCM	Values
1.	Molar mass	380
2.	Melting temperature (°C)	60
3.	Density (kg/m³)	778
4.	Latent heat (kJ/kg)	189.47
5.	Thermal conductivity (W/m K)	0.231
6.	Viscosity (mPa-s)	9.27

(a)

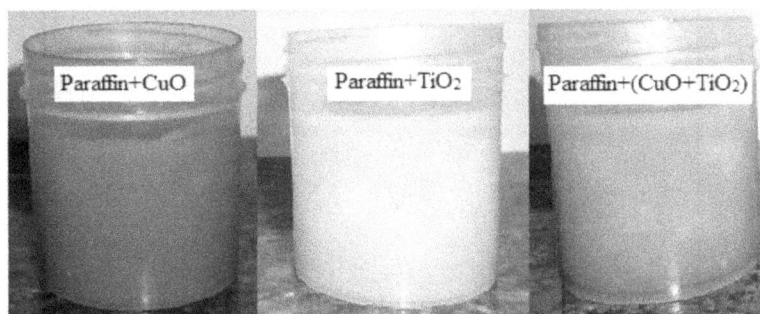

(b)

FIGURE 7.4 Sedimentation photographs of the composite PCMs in (a) melted stage (b) solidified stage[3].

from Sigma-Aldrich, USA. Surfactant mass fraction was 1.2 times that of nanomaterial mass fraction during the preparation of all composite PCMs. Afterwards, they were subjected to 60, 70, 80, and 90 minutes of ultrasonic vibration at a frequency of 40 kHz, with a corresponding amount of 0.25, 0.5, 0.75, and 1.0 wt%. To maintain the paraffin in a liquid state, the vibrator temperature was held above the melting point of the paraffin during the creation of composite PCMs. Figure 7.4 shows digital photos of the suspension of the composite PCMs with 1.0 wt% CuO, TiO_2, and hybrid nanomaterials prior to centrifugation.

7.2.3 CHEMICAL STABILITY OF COMPOSITE PCMs

Figure 7.5 shows the IR spectra of paraffin and a PCM composite. There are symmetrical stretching vibration peaks of CH_2 in paraffin at 2917.51 cm^{-1} and 2848.52 cm^{-1}. The bending vibration maxima at 1463.04 cm^{-1} and 719.82 cm^{-1} are due to the CH_3 group in paraffin. When FTIR and composite PCM spectra show no shift in absorption peaks, it is clear that the spectrum is paraffin. The FTIR results demonstrate that the composite PCM containing 1.0wt% hybrid nanomaterials has no additional peaks in the spectra when compared to paraffin, and the spectra seem to be same. This indicates that the paraffin and hybrid NPs are merely physically interacting, and there is no chemical reaction between them[3, 4].

7.2.4 INFLUENCE OF NPs ON THE THERMAL CONDUCTIVITY

The LFA 447 Nano Flash analyzer was used to assess the thermal conductivity of the composite PCMs comprising 0.25, 0.5, 0.75, and 1.0wt% nanomaterials, as the thermal conductivity of the PCMs is critical in determining the energy storage and

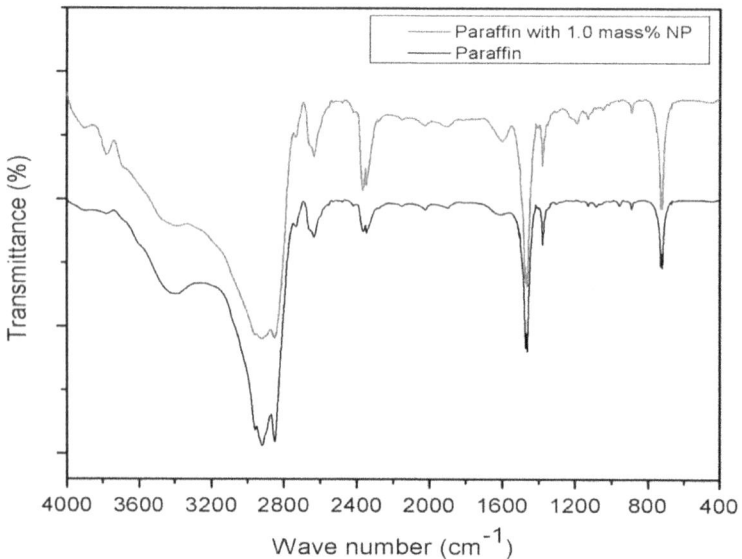

FIGURE 7.5 FTIR spectra of the paraffin and composite PCM with 1.0wt% hybrid nanomaterials[3].

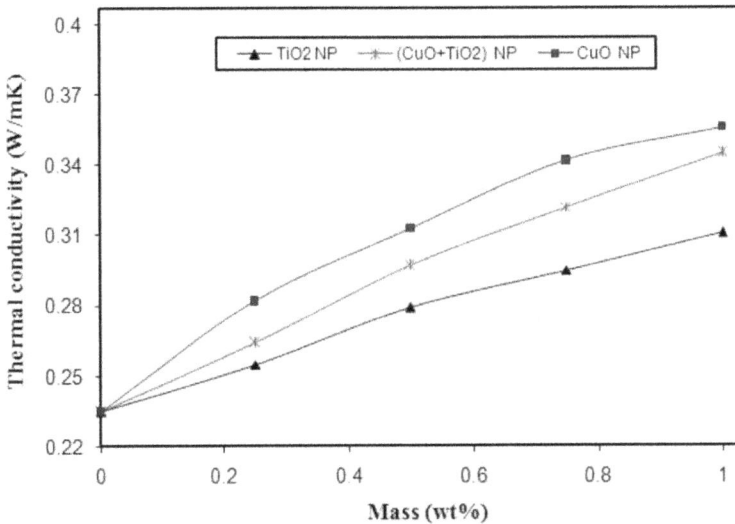

FIGURE 7.6 Influence of mass fraction of hybrid nanomaterials on the thermal conductivity[3].

release rates. CuO and TiO$_2$ NPs were also added to paraffin, and the thermal conductivities of the mixtures were measured to determine the degree of improvement for the various concentrations. Figure 7.6 shows that the thermal conductivity of paraffin with CuO, TiO$_2$, and the nano-mixture increases as the concentration of nanomaterials increases. This finding shows that the concentration of nanomaterials is responsible for the increased thermal conductivity of the composite PCMs. It was also revealed that the improvement in paraffin quality achieved with CuO was superior to that obtained with TiO$_2$ and the nano-mixture. The thermal conductivity of the CuO NPs is higher than that of the TiO$_2$ NPs and nano-mixture. Furthermore, it is widely known that cylindrical NPs have a higher thermal conductivity than spherical NP. Furthermore, the greater heat transfer surface area of the particles compared to their spherical counterparts means that the particles will transfer heat more effectively. SEM investigations confirmed that the morphologies of CuO and TiO$_2$ NP samples were rods and spheres. The base material's thermal conductivity will improve as a result of a better aspect ratio, which increases the surface area[5, 6]. CuO, TiO$_2$, and hybrid nanomaterials enhanced the paraffin's thermal conductivity by 51.49%, 32.34%, and 46.81% compared to pure paraffin. An increase in the nanomaterial content over 1.0% of the base material might shorten the distance between the particles, leading to a greater degree of interaction and eventually, the formation of agglomerates. Therefore, this could possibly result in an increased melting and freezing time, which would decrease thermal conductivity. It is therefore vital to find the best concentration of nanomaterials to improve the heat transfer efficiency of the LTES system.

7.2.5   EFFECT OF NPS ON THE VISCOSITY

It was discovered that varying mass fractions of CuO, TiO_2, and hybrid nanomaterials affected paraffin viscosity when added to paraffin in varying amounts. Figure 7.7 shows that the viscosity of the composite PCMs increased in a nonlinear approach as the amount of nanomaterials in the composite PCMs increased. As well, the viscosity of paraffin increased with CuO more than with TiO_2 and hybrid nanomaterials. Increased viscosity results from a variety of parameters, including the density of the nanomaterials, particle size and shape, pH, and the manner in which they are disseminated. TiO_2 viscosity improvement is also less affected by the rod form of TiO_2 than CuO. The viscosity of paraffin was raised by 7.76, 4.85, and 6.15 using a 1.0 wt% mass fraction of CuO, TiO_2, and hybrid nanomaterials.

7.2.6   THERMOGRAVIMETRY ANALYSIS OF COMPOSITE PCMS

Figure 7.8 depicts the TG curves for paraffin and PCM composites with 1.0wt% hybrid nanomaterials. When paraffin and composite PCM materials are combined, only one step of weight is lost. The composite PCM's weight loss is less than that of the paraffin, and this is because the hybrid nanoparticles are more widely dispersed. Figure 7.8 shows that the paraffin molecular chains break down at temperatures between 210°C and 350°C during the process' first stage. The paraffin's hybrid nanomaterials have worked as heat retardants, thereby delaying the paraffin's dissolution. Explosive molecules could be prevented from entering the gas phase by using this retardant. The improved thermal stability of the composite PCM over that of pure paraffin is shown in this result.

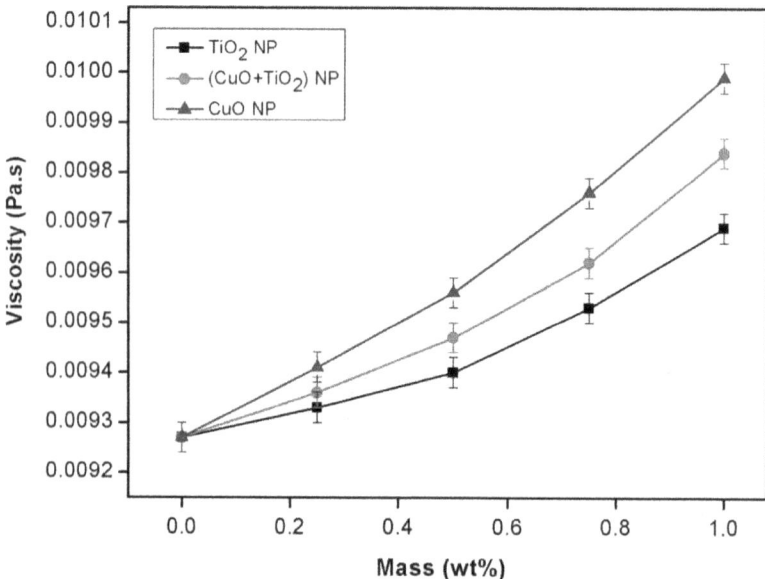

FIGURE 7.7   Effect of hybrid nanomaterials on the viscosity[3].

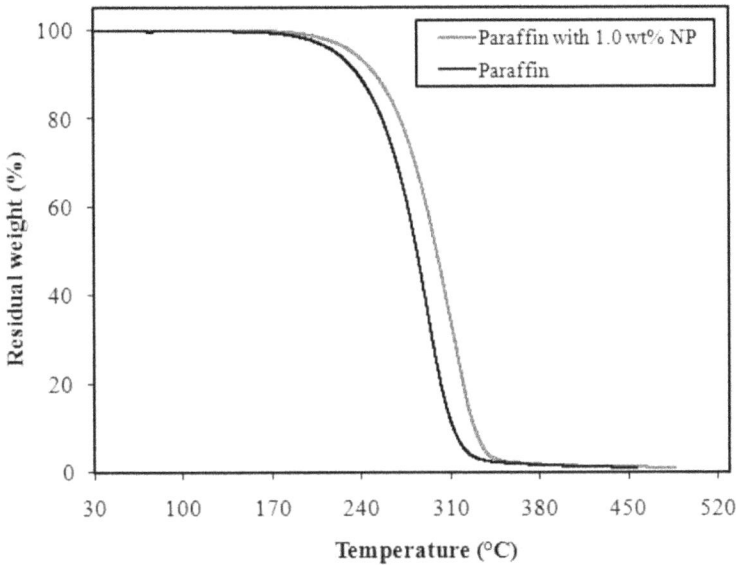

FIGURE 7.8   TG curves of the paraffin and composite PCM with 1.0wt% hybrid nanomaterials[3].

7.3   PHASE CHANGE PROPERTIES OF COMPOSITE PCMs

DSC was used to examine the phase change temperatures and latent heats of solid and liquid PCMs. It focuses on the effects of different mass fractions of hybrid nano-materials on the melting and solidification points. Also, the number of thermal cycles had an impact on the temperature and latent heat changes in composite PCMs.

7.3.1   THERMAL PROPERTIES OF COMPOSITE PCMs

Hybrid nanomaterials embedded in the base material have been found to influence the phase change temperatures and latent heats of the base material and composite PCMs, as shown in Figure 7.9. Melting temperatures ranged from 54°C to 62°C, suggesting that the base material and composite PCMs had departed from solid to liquid.

The PCM thermal characteristics in respect to the mass fraction of hybrid nano-materials are shown in Table 7.2. Composite PCMs melt and freeze at higher temper-atures than basic materials. For the melting and freezing temperatures, the highest variations are −0.87% and −1.02%, respectively, and these might be neglected. Composite PCMs have lower melting and freezing latent heat than the base material, with maximum changes of 3.56% and 3.82%, respectively.

Since the energy storage and release rates of LTES systems are hardly affected by the changes, the system claims that they will be negligible. While base materials will work, these minor cuts may allow the composite PCMs to be more efficient at storing and releasing energy.

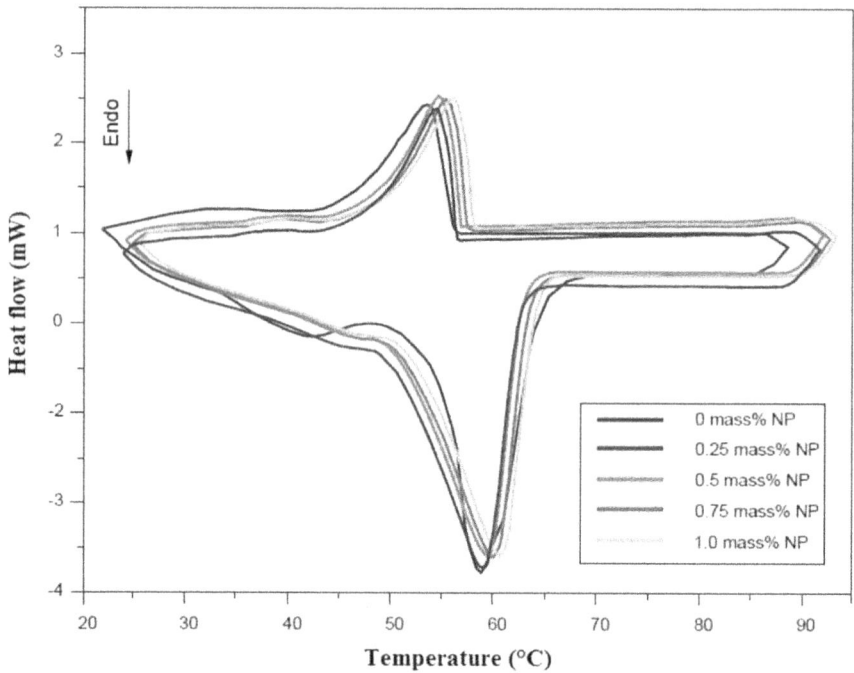

FIGURE 7.9 DSC measurements for the paraffin and composite PCMs[3].

TABLE 7.2
**Composite PCM Thermal Characteristics as a Function of Hybrid
Nanomaterial Mass Fraction**[3]

Mass Fraction of Hybrid Nanomaterials (wt%)	Melting Temperature (°C)	Freezing Temperature (°C)	Melting Latent Heat (kJ/kg)	Freezing Latent Heat (kJ/kg)
0	60.23	56.47	197.62	189.47
0.25	60.34	56.54	196.22	188.06
0.5	60.47	56.69	194.68	186.23
0.75	60.63	56.84	192.34	184.89
1.0	60.84	56.96	190.06	182.73

7.3.2 THERMAL RELIABILITY OF COMPOSITE PCMs

Table 7.3 shows that the phase change temperatures and latent heats in all thermal cycles were small: −0.72% and −0.84% were the largest disparities in melting and freezing temperatures. The most reduction in latent heat melting and freezing

TABLE 7.3
Thermal Properties of Composite PCMs with Respect to Number of Thermal Cycles

No. of Thermal Cycles	Melting Temperature (°C)	Freezing Temperature (°C)	Melting Latent Heat (kJ/kg)	Freezing Latent Heat (kJ/kg)
0	60.84	56.96	190.06	182.73
500	60.99	57.09	189.27	181.27
1000	61.15	57.22	188.34	180.78
1500	61.06	57.31	187.18	179.24
2000	61.28	57.42	186.59	178.57

was 2.27% respectively and 1.83%. The variations were caused by changes in the composite PCM's thermal and physical properties. These findings may be ignored because the additional thermal cycles had no discernible effect on the system's performance.

7.4 PHASE CHANGE BEHAVIOR OF COMPOSITE PCMs

Thermal and heat transfer properties of the composite PCMs were investigated experimentally while they solidified and melted.

7.4.1 EXPERIMENTAL SETUP

It was employed as a heat transfer fluid to enable temperature adjustment of the water from 20°C to 90°C. The HTF was kept at a constant 35°C temperature and pumped into the TES tank during the cooling operation. In addition, the HTF was heated to 80°C and pumped into the TES tank by the use of a pump. The constant temperature tank was kept stable by having temperature controllers with an electrical stirrer inside of it. The Agilent 34972A data logger (USA) was used to measure the PCM temperature with a 5-sec time interval, and a thermocouple was positioned at the center of the spherical encapsulation. The experiment was repeated five times to guarantee that the measurement data was reproducible.

7.4.2 SOLIDIFICATION AND MELTING CHARACTERISTICS

The melting and freezing rates during heating and cooling modes of operation can identify composite PCMs with improved thermal conductivity. The base material (paraffin) and composite PCMs have a typical temperature profile displayed in Figure 7.10. The base material and composites' initial temperature was 37°C. The PCMs' temperature gradually increased throughout the procedure. A temperature

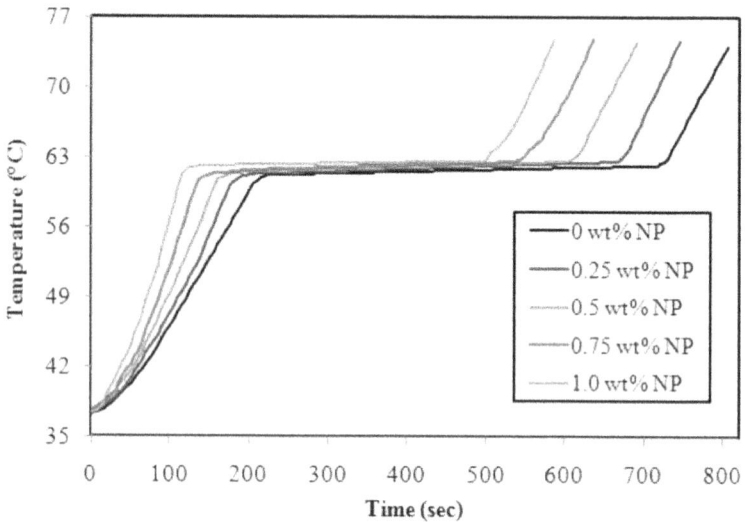

FIGURE 7.10 Melting processes of the paraffin and composite PCMs[3].

increase of 60°C would turn the PCMs liquid, storing the heat energy they had previously absorbed through their solid state. As the HTF and PCMs became more similar in temperature, the advancement of the process continued. It is shown in Figure 7.10 that the melting times of the PCMs with 0, 0.25–0.75, 0.75, and 1.0wt% hybrid nanomaterials were 705s, 655 s, 605 s, 504 s, and 495s. Thermal conductivity and convective flow improvements have been made possible by using nanoparticles as a base material (composite PCM) for the enhanced heat transfer rate. It took 5.67, 14.18, 23.40, and 29.78% less time to melt the composite PCMs containing 0.25, 0.5, 0.75, and 1.0wt% hybrid nanomaterials than it did to melt the base material[7, 8].

Figure 7.11 shows the basic material and composite PCMs were kept at 72°C during the initial stages of freezing. The temperature of the PCMs dropped as the freezing process continued, and the base material and composite PCMs began to freeze as the process continued. PCMs tend to shift between the liquid and solid states. There was no additional temperature decrease in the PCMs as they were freezing or solidifying. Once the PCMs were completely frozen, their subsequently rationale would be to cool to the HTF's temperature.

There were four different times necessary to freeze the PCMs with different amounts of hybrid nanomaterials: 1340, 1260, 1160, and 955 seconds. Composite PCMs with 0.25, 0.5, 0.75, and 1.0 wt% hybrid nanomaterials had freezing times that were 5.97, 13.43, 20.15, and 28.73% shorter than base materials. Furthermore, it is obvious that enhanced heat transfer properties have been attained due to the incorporation of NPs in the base material. Nanomaterials are widely used in PCMs, due to the fact that freezing process resulted in increased conduction. An increase in

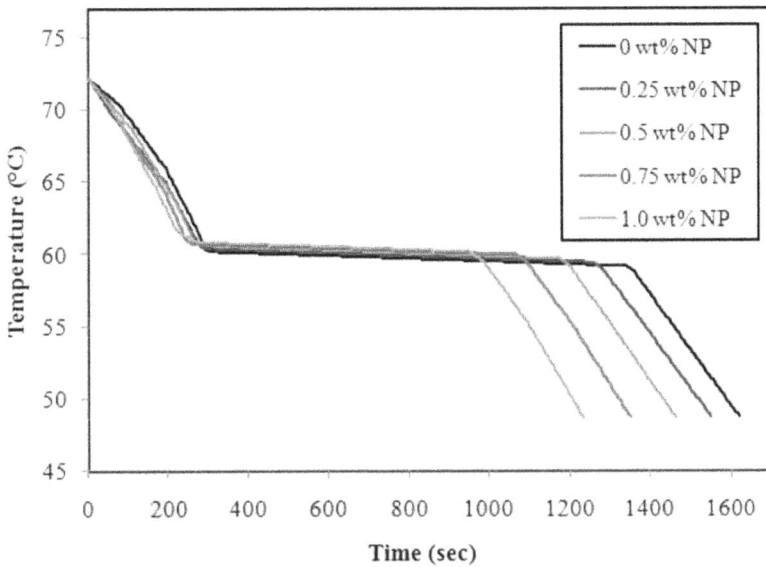

FIGURE 7.11   Freezing processes of the paraffin and composite PCMs[3].

thermal conductivity was indirectly confirmed by the composite PCMs' enhanced ability to transfer heat during melting and freezing.

7.5   COMPOSITE PCMs WITH DIFFERENT TYPES OF NANOPARTICLES

Composite PCM was developed by combining 70:30 blends of lauric acid (LA) and stearic acids (SA). The thermal conductivity of LA/SA mixtures was enhanced with the addition of CuO, ZnO, and TiO_2 NPs with 1.0 wt% content. When heated and cooled, the thermal conductivity and heat transfer properties of composite PCMs were examined.

7.5.1   CHARACTERIZATION OF THE NPs

The morphology of CuO, ZnO, and TiO_2 generated particles was examined using field emission scanning electron microscopy (FESEM, LEO 1530, Zeiss, Germany). Figure 7.12 shows the different morphologies of TiO_2, ZnO, and CuO NPs, which are all spherical, cylindrical, and rod-like, respectively. Heat transfer would be more efficient with the rod and spherical shapes, which offer varied surface areas. The FESEM pictures also reveal the presence of NP aggregates. The assembly of NPs can be controlled by sonication during the creation of the composite PCMs[10, 11].

(a)

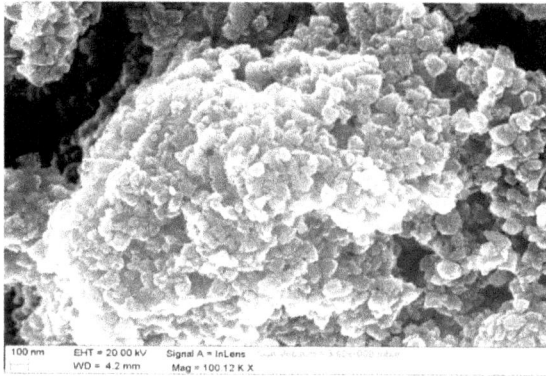

(b)

(c)

FIGURE 7.12 SEM images of (a) TiO_2 NPs, (b) ZnO NPs, (c) CuO NPs[9].

7.5.2 PREPARATION OF COMPOSITE PCMS

Lauric acid (43–45°C) and stearic acid (54°C) were obtained from Alfa Aesar and Thermofisher Scientific (India) Private Limited, respectively. With 70 g of lauric acid and 30 g of stearic acid, this base material (LA/SA) was used to make composite PCMs. The thermophysical characteristics of the LA/SA mixture are presented in Table 7.4. NPs consisting of 1.0 wt% TiO_2, ZnO, and CuO were considered as supportive materials with a total mass fraction of 1.0 wt%. The two-step method was used to distribute these types of NPs in the LA/SA mixture.

Composite PCM preparation needs uniform dispersion, no particle clustering, and no chemical reaction between the NPs and the base material. SDBS (as a capping agent) was added in the LA/SA mix at a concentration of 1.2 wt% to enhance the melting and solidification rates of the composite PCMs. In order to increase thermal conductivity, this was done by incorporating NPs into the mix in a constant manner.

Ultrasonic dispersion of the NPs in the base material was recommended to ensure a stable dispersion. NPs were suspended in the base material using a high-frequency vibrating device called a sonicator (40 kHz). A 60-minute residing period is predicted for the TiO_2, ZnO, and CuO NP mass fraction in the composite PCMs. The temperature of the LA/SA combination was kept above the melting point during the sonication process. Figures 7.13 and 7.14 show the prepared composite PCMs' digital picture and chemical structure are presented.

TABLE 7.4
Thermophysical Properties of the LA-SA Mixture[9]

S.No	Properties of the PCM	Values
1.	Melting temperature (°C)	31–34
2.	Density (kg/m³)	921
3.	Latent heat (kJ/kg)	174.68
4.	Thermal conductivity (W/m K)	0.198
5.	Viscosity (mPa-s)	13.352

FIGURE 7.13 Digital images of the base material and composite PCMs[1].

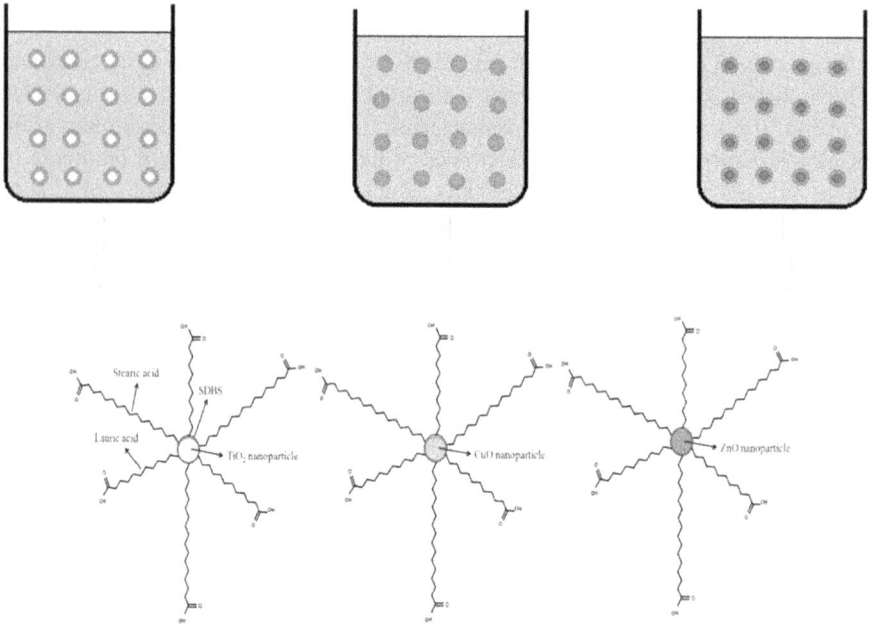

FIGURE 7.14   Chemical structure of the composite PCMs[9].

7.5.3   Influence of NPs on the Thermal Conductivity

The ability of composite PCMs to store and release energy is determined by the heat transfer rate and the temperature difference between the HTF and PCM. Because of their high thermal conductivity, PCM melting and solidification occur more rapidly. Furthermore, PCM heat conductivity improvements would dramatically decrease the time required for the PCMs to melt and for solidification. Thermal conductivity was expected to increase when the base material was divided into different types of NPs.

Figure 7.15 clearly demonstrates that the thermal conductivity of the composite PCMs was raised and estimated at 34.85%, 46.97%, and 62.12%, respectively, for 1.0wt% of TiO_2, ZnO, and CuO, for all NPs at 0%. Composite PCM with CuO NPs outperformed PCMs with TiO_2 and ZnO NPs in terms of enhancement potential. In contrast to the TiO_2 and ZnO NP spherical shapes, the CuO NPs' s surface area and length were greater than those of the synthesized rod-shaped particles, which limited their dispersion. Composite PCMs have improved thermal conductivity because of the different shapes, sizes, and compositions of the NPs used. Composite PCM materials, which contain CuO nanoparticles, are more suitable for improving LTES system performance than those containing TiO_2 and ZnO NPs.

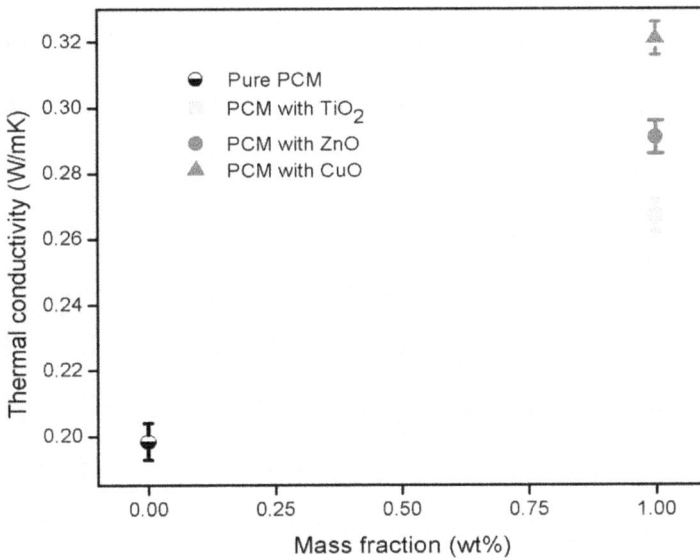

FIGURE 7.15 Thermal conductivity of the base material and composite PCMs[9].

7.5.4 EFFECT OF NPS ON THE VISCOSITY

Due to the increased viscosity caused by the NPs in the base material, it is necessary to assess the composite PCMs' overall viscosity. Additionally, the increased viscosity that arises from the addition of NPs to the base material can be tolerable to a certain amount. These increases could have a negative impact on LTES system performance since they will make composite PCMs less effective. Figure 7.16 depicts the effect of the NPs on the composite PCMs' viscosity.

In the experiments, the composite PCMs' viscosity increased, and the results also show that the composite PCMs have a maximum increase in viscosity than the basic material. ZnO, TiO$_2$, and CuO were exposed to comprise 2.59%, 2.94%, and 3.67%, respectively. It is also noticed that 1.0wt% CuO composite PCM has shown better performance enhancements when compared to 1.0wt% TiO$_2$ and ZnO composite PCMs. Also, the improved viscosity of the composite PCMs will have no impact on the LTES systems' heat transfer characteristics. Nanofluids, which flow in a pipe line, are different from the composite PCMs in the spherical encapsulation of the LTES system. Because the viscosity augmentation of the PCMs does not contribute any thermal resistance to heat transfer, it could be ignored[12, 13].

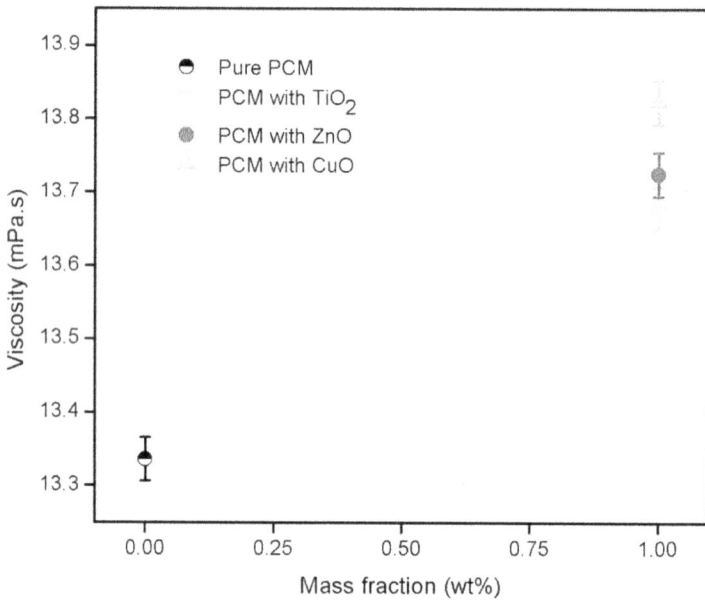

FIGURE 7.16 Viscosity of the base material and composite PCMs[9].

7.5.5 THERMOGRAVIMETRY ANALYSIS OF COMPOSITE PCMS

Base material and composite PCMs were examined for stability using TG testing. A comparison of base material and composite PCM weight decrease is shown in Figure 7.17. The base material lost weight when heated to temperatures between

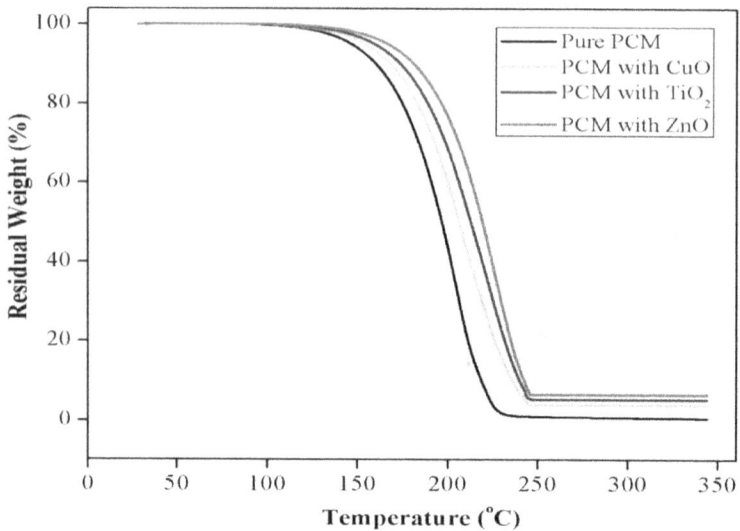

FIGURE 7.17 TG curves of the base material and composite PCMs[9].

110°C and 230°C. The weight loss at 120°C was about 6%, which was contributed to by the reduction of absorbed water. Degradation of the polymer material occurred at temperatures between 200°C and 230°C, with monomer fragmentation being the predominant cause. In comparison with their base materials, composite PCMs lose weight more rapidly between 120°C and 250°C than composite PCMs.

For composite PCMs made of CuO, TiO$_2$, and ZnO, the maximum weight loss was found to be at 250°C, with values of 95%, 93%, and 91%, respectively. Because the NP additions in the base material have acted as a barrier against temperature degradation, it is reasonable to presume that the base material would have deteriorated even more if the NP additives had not been included. NPs in the base material have better thermal stability, which is evident because the base material alone is inferior to the modified material. Furthermore, PCM thermal stability is highly dependent on the types of NPs distributed. The heat storage capacity of composite PCMs for building heating applications is tested in the 23°C to 48°C temperature range. Figure 7.17 shows no weight loss for this temperature range. As a result, the composite PCMs which are made for building heating applications are able to perform without decomposition.

7.6 PHASE CHANGE PROPERTIES OF COMPOSITE PCMs

DSC analysis was used to estimate composite PCM phase change temperatures and latent heats of solidification and melting, as well as the impact of adding NPs (Figure 7.18). Composite PCM phase change temperatures and latent heats were also determined using thermal cycles.

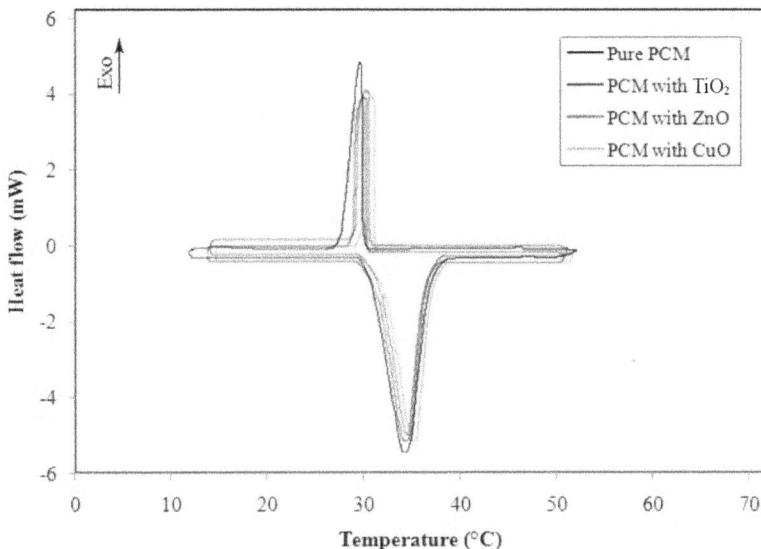

FIGURE 7.18 DSC measurements of the base material and composite PCMs with TiO$_2$, ZnO, and CuO NPs[9].

7.6.1 THERMAL PROPERTIES OF COMPOSITE PCMs

NPs were introduced to composite PCMs, and DSC measurements were performed to determine the melting, solidification, and latent heat of the new materials. Table 7.5 displays the solidification temperature fluctuation in accordance with the addition of different kinds of NPs. The composite PCMs comprising 1.0 wt% TiO_2, ZnO, and CuO NPs had solidification temperatures that varied by −0.68, −0.99, and −1.19% from those of the base material.

Table 7.5 shows the melting point changes when different types of NPs are added. The melting points of TiO_2 and ZnO composites differ from the base material by −1.11% and −1.38%, respectively. This slight variation in solidification and melting temperatures can be disregarded because it would not influence the LTES system.

Table 7.5 depicts the latent heat changes during solidification and melting, respectively. As compared to the base material, the latent heat of solidification was reduced by 2.06%, 2.35%, and 2.57% when using 1.0 wt% composite PCMs containing TiO_2, ZnO, and CuO NPs. For the composite PCMs containing 1.0 wt% TiO_2, ZnO, and CuO NPs, the most significant decreases in melting latent heat were 2.12%, 1.89%, and 1.76%, respectively. Insignificant changes in the latent temperatures of solidification and melting can have no impact on the energy storage and release capabilities of the composite PCMs.

7.6.2 THERMAL RELIABILITY OF COMPOSITE PCMs

PCMs in LTES systems are subjected to more melting and solidification cycles, resulting in longer PCM lifespans. The thermal stability of NPs made with a base material and having 1.0 wt% TiO_2, ZnO, and CuO was investigated for 5000 heat cycles. A composite PCM phase change temperature variation owing to thermal cycles is depicted in Figure 7.19. For 1.0wt% TiO_2, ZnO, and CuO NPs, the maximum variations in solidification temperatures were −2.08%, −1.55%, −1.39%, and −1.24%[14, 15]. Figure 7.20 shows that for 1.0 wt% TiO_2, ZnO, and CuO NPs, the highest melting temperature changes were −2.34%, −1.78%, −2.12%, and −1.90%, respectively. Similar to this, the 201.55%, −1.39%, and −1.24% changes in melting temperatures were observed for 1.0 wt% TiO_2, ZnO, and CuO NPs, respectively. The fluctuations in phase change temperatures, when examined from the standpoint of a

---

TABLE 7.5
Thermal Properties of Composite PCMs with 1.0wt% of Different NPs[9]

Name of the PCMs	Melting Temperature (°C)	Solidification Temperature (°C)	Melting Latent Heat (kJ/kg)	Solidification Latent Heat (kJ/kg)
(LA+SA)	34.09	29.27	176.98	174.68
(LA+SA)+TiO_2	34.38	29.47	173.22	170.19
(LA+SA)+ZnO	34.47	29.56	173.64	170.57
(LA+SA)+CuO	34.56	29.62	173.86	171.08

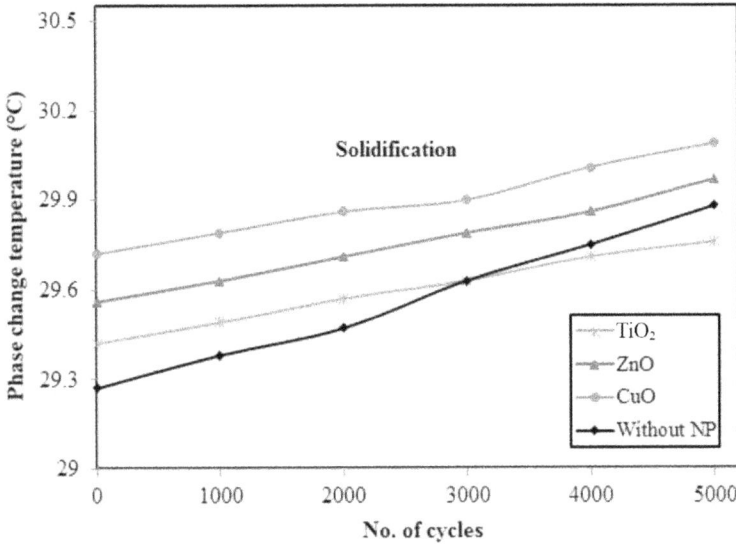

FIGURE 7.19 Effect of thermal cycles on phase change temperature during solidification[9].

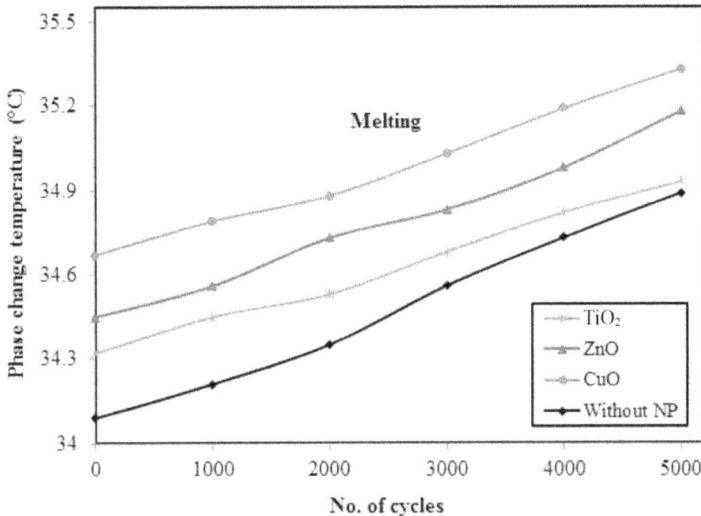

FIGURE 7.20 Effect of thermal cycles on phase change temperature during melting[9].

space heating system, could be ignored. Figure 7.21 shows the different variations in solidification latent heat for composite PCMs based on thermal cycles. Solidification latent heat decreases of 2.56%, 1.54%, 1.634%, and 1.484% for 1.0 wt% TiO_2, ZnO, and CuO NPs have been found to be possible. Figure 7.22 shows that for 1.0 wt% TiO_2, ZnO, and CuO NPs, the melting latent heat temperatures were lowered by

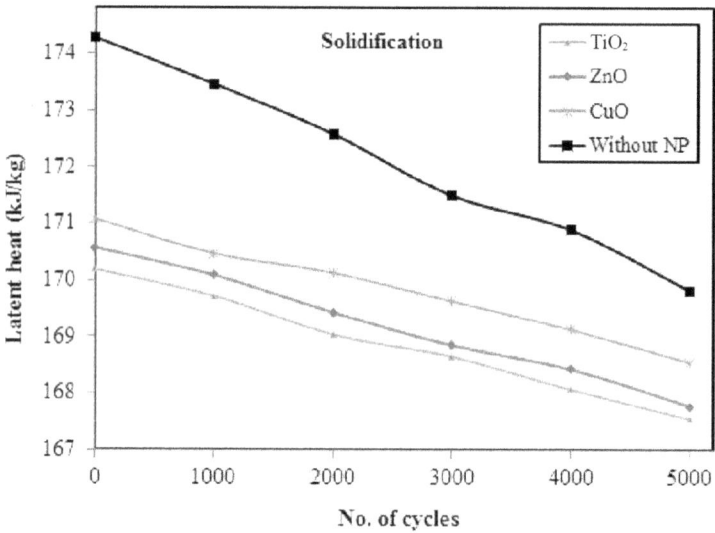

FIGURE 7.21   Effect of thermal cycles on latent heat during solidification[9].

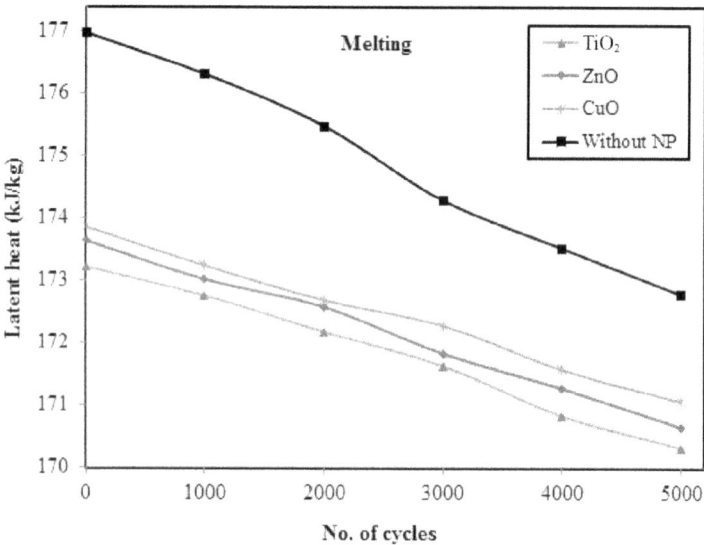

FIGURE 7.22   Effect of thermal cycles on latent heat during melting[9].

2.37%, 1.67%, 1.72%, and 1.6%, respectively. The latent heat reductions for solidification and melting are small for LTES systems in comparison. Phase change temperatures and latent heats of composite PCMs differ slightly as a result of heat-related
physical changes inside the composite PCMs. The composite PCMs were proven to

be more thermally reliable than the base material. Composite PCMs are projected to have significantly reduced latent heat storage/release capabilities when subjected to more than 5000 thermal cycles.

7.7 PHASE CHANGE BEHAVIOR OF COMPOSITE PCMs

Thermal and heat transfer properties of the composite PCMs were studied experimentally during cooling and heating processes.

7.7.1 EXPERIMENTAL SETUP

Water served as the heat transfer fluid, and its temperature range was from 10°C to 85°C. With a pump circulating the chilled HTF, the temperature was maintained at 15°C.

A pump was used to move HTF from the heater to the TES tank while keeping the temperature constant at 50°C. When test conditions insisted, two tanks were each equipped with an electrical stirrer. The temperature of the PCMs was monitored every 5 seconds using a thermocouple associated to the data recorder. A data logger (Agilent 34972A model, USA) was utilized to monitor the testing conditions constantly with the thermocouples put in both tanks. To ensure the repeatability of the measurement data, the tests were run five times.

7.7.2 SOLIDIFICATION AND MELTING CHARACTERISTICS

During the melting process, the temperature range of the base material and the composite PCM can be shown in Figure 7.23. The initial stages of the melting process were monitored, when the PCMs had a temperature of 23°C. When enough time had passed, the base material and the PCM composites had warmed up to the point where the PCMs had dissolved. After melting, PCMs totally switched to a liquid phase until further temperature increases. The heat energy stored in the PCMs, as provided by the HTF, is clearly evident. As time passed, the PCMs' temperatures rose to liquid state after becoming solid, with the implication that no further heat energy was stored. The base material and the various PCMs' complete melting times are 965 s, 855 s, 815 s, and 760 s, respectively, as shown in Figure 7.23. The TiO_2, ZnO, and CuO in the composite PCMs saved 11.39% of the entire melting time, whereas the base material required 15.54 and 21.24% to melt completely. The conductive and convective heat transfer was enabled by the dispersed NPs in the base material, which accelerated the melting process. From Figure 7.24, it is clear that the base material and composite PCMs containing 1.0wt% TiO_2, ZnO, and CuO NPs required solidification times of 1260 s, 1190 s, and 1085 s. Overall, when compared to the base material, the solidification time savings for composite PCMs with 1.0wt% TiO_2, ZnO, and CuO is 5.56%, 13.89%, and 19.84%. During the solidification process, the NPs in the base material are aided in the conductivity and convection of heat transfer. The NPs aid in the melting and solidification processes by accelerating the passage of heat conduction[16].

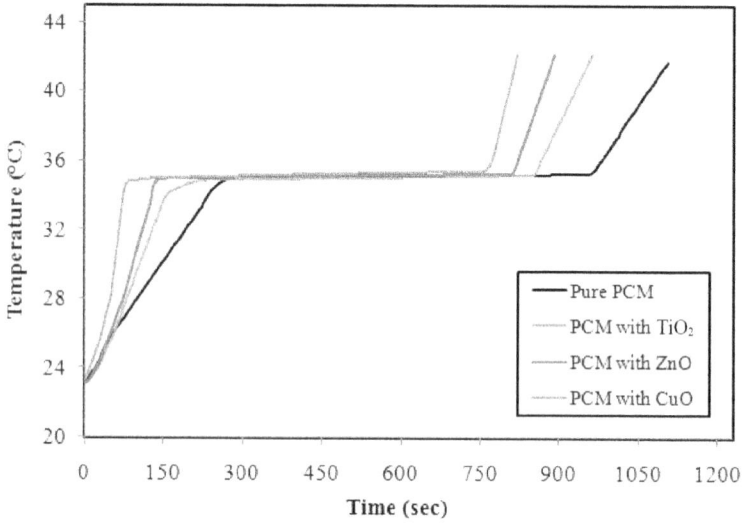

FIGURE 7.23   Melting processes of the base material and composite PCMs[9].

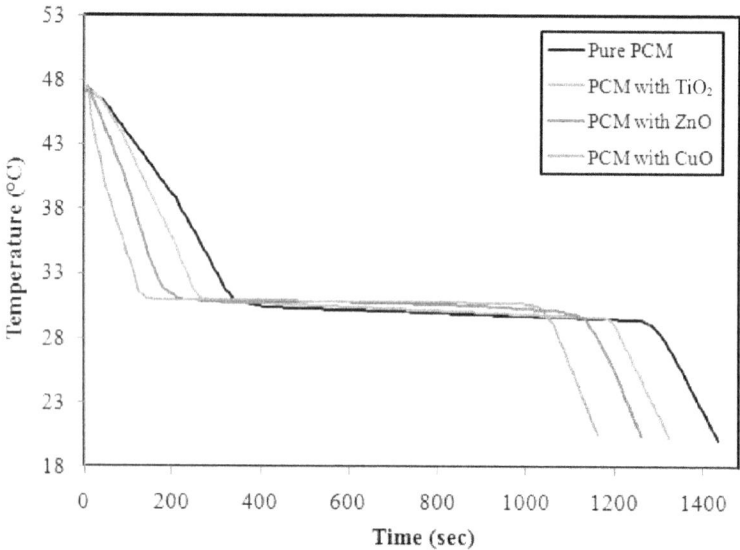

FIGURE 7.24   Solidification processes of the base material and composite PCMs[9].

7.8   SUMMARY

In this chapter, the experimental investigation of two types of composite PCMs was undertaken to determine their potential for use in low-temperature heat thermal storage. For the preparation of composite PCMs, paraffin as the base material and hybrid nanomaterials in different mass fractions were used. Composites made of LA/SA mix (70:30 by weight) served as a base material for the PCMs while TiO_2, Zn, and CuO NPs (1.0% by weight) provided support. The surfactant SDBS was utilized in both the composite PCMs to maintain the nanomaterials' consistency in the base materials. Both types of composite PCM using NPs and measuring the number of thermal cycles showed little divergence from the base material in terms of phase change temperatures and latent heat. In terms of latent thermal energy storage devices, these little differences would not impact their performances.

The first study determined the thermal conductivity of various composite PCMs, each made up of a different amount of mass fraction CuO, TiO_2, and hybrid nanomaterials. The results showed that the composite PCMs using hybrid nanomaterials had an enhancement value between the CuO and TiO_2 NP-based composite PCMs. PCM thermal conductivity with 1% TiO_2 and ZnO NPs was higher than PCM with 1 wt% CuO NPs.

Both composite PCMs, using CuO and TiO_2, were evaluated and the viscosity of the nanomaterials was found to be the highest in the composite PCMs with CuO. However, a different study found that a mixture of TiO_2 and ZnO NPs added to composite PCM and resulted in a higher viscosity than a similar mixture of TiO_2 and ZnO NPs. Finally, the results of the heating and cooling tests show that the melting and solidification times for both the composite PCMs and their base materials have been significantly lowered.

REFERENCES

[1] Baghbanzadeh, M., Rashidi, A., Rashtchian, D., Lofti, R. & Amrollahi, A., 'Synthesis of spherical silica/multiwall carbon nanotubes hybrid nanostructures and investigation of thermal conductivity of related nanofluids', *Thermochimica Acta*, vol. 549, pp. 87–94, 2012.

[2] Harikrishnan, S. & Kalaiselvam, S., 'Preparation and thermal characteristics of CuO-oleic acid nanofluids as a phase change material', *Thermochimica Acta*, vol. 533, pp. 46–55, 2012.

[3] Harikrishnan, S., Deepak, K. & Kalaiselvam, S., 'Thermal energy storage behavior of composite using hybrid nanomaterials as PCM for solar heating systems', *Journal of Thermal Analysis and Calorimetry*, vol. 115, no. 2, pp. 1563–1571, 2014.

[4] He, Q., Wang, S., Tang, M. & Liu, Y., 'Experimental study on thermophysical properties of nanofluids as phase-change material (PCM) in low temperature cool storage', *Energy Conversion and Management*, vol. 64, pp. 199–205, 2012.

[5] Jana, S., Salehi-Khojin, A. & Zhong, W.-H., 'Enhancement of fluid thermal conductivity by the addition of single and hybrid nano-additives', *Thermochimica Acta*, vol. 462, pp. 45–55, 2007.

[6] Kenisarin, M. & Mahkamov, K., 'Solar energy storage using phase change materials', *Renewable and Sustainable Energy Reviews*, vol. 11, pp. 1913–1965, 2007.

[7] Mehrali, M., Latibari, S. T., Mehrali, M., Metselaar, H. S. C. & Silakhori, M., 'Shape-stabilized phase change materials with high thermal conductivity based on paraffin/graphene oxide composite', *Energy Conversion and Management*, vol. 67, pp. 275–282, 2013.

[8] Murshed, S. M. S., Leong, K. C. & Yang, C., 'Enhanced thermal conductivity of TiO_2-water based nanofluids', *International Journal of Thermal Sciences*, vol. 44, pp. 367–373, 2005.

[9] Harikrishnan, S., Deenadhayalan, M. & Kalaiselvam, S., 'Experimental investigation of solidification and melting characteristics of composite PCMs for building heating application', *Energy Conversion and Management*, vol. 86, pp. 864–872, 2014.

[10] Ozerinc, S., Kakac, S. & Yazicioglu, A. G., 'Enhanced thermal conductivity of nanofluids: A state-of-the-art review', *Microfluidics and Nanofluidics*, vol. 8, pp. 145–170, 2010.

[11] Timofeeva, E. V., Routbort, J. L. & Singh, D., 'Particle shape effects on thermophysical properties of alumina nanofluids', *Journal of Applied Physics*, vol. 106, no. 1, p. 014304, 2009.

[12] Timofeeva, E. V., Yu, W., France, D. M., Singh, D. & Routbort, J. L., 'Base fluid and temperature effects on the heat transfer characteristics of SiC in ethylene glycol/H$_2$O and H$_2$O nanofluids', *Journal of Applied Physics*, vol. 109, p. 014914, 2011.

[13] Wang, J., Xie, H. & Xin, Z., 'Thermal properties of paraffin based composites containing multi-walled carbon nanotubes', *Thermochimica Acta*, vol. 488, pp. 39–42, 2009.

[14] Wu, S., Zhu, D., Zhang, X. & Huang, J., 'Preparation and melting/freezing characteristics of Cu/Paraffin nanofluids as phase-change material (PCM)', *Energy and Fuels*, vol. 24, pp. 1894–1898, 2010.

[15] Xie, H., Wang, J., Xi, T. & Liu, Y., 'Thermal conductivity of suspensions containing nanosized SiC particles', *International Journal of Thermophysics*, vol. 23, no. 2, pp. 571–580, 2002.

[16] Zeng, J. L., Cao, Z., Yang, D. W., Xu, F., Sun, L. X., Zhang, X. F. & Zhang, L., 'Effects of MWNTs on phase change enthalpy and thermal conductivity of a solid-liquid organic PCM', *Journal of Thermal Analysis and Calorimetry*, vol. 2, pp. 507–512, 2009.

8 Nanofluids

CONTENTS

8.1 INTRODUCTION

A single component, a mixture, or a combination of two or more components can make up the nanomaterials scattered in a fluid. The amount of heat carried over by the nanofluid is reliant on the nanomaterials distributed, which ultimately depends on its basic properties at the nanoscale. The performance of a nanofluid is dependent on the stability and dispersion of nanoparticles in the system, which is promising in many applications. While ultrasonication and surfactants are employed to ensure correct dispersion of nanofluids, pH management is also critical to their long-term stability[1].

8.2 PRODUCTION OF NANOPARTICLES AND NANOFLUIDS

Using current fabrication methods, it is now possible to manufacture materials with a nanometer-scale structure. In comparison to bigger (micron size and larger) particles of the same substance, those in NPs have different physical and chemical properties. Physical and chemical processes are the two primary methods for fabricating NPs, which have been utilized to make nanofluids from a wide range of materials[2].

There are many types of NP materials used in electronics, including oxide ceramics (Al_2O_3), nitrides (SiN), carbides (SiC), metals (Ag, Cu), semiconductors (TiO_2), and composites including NP core polymer shell composites. Material NPs were synthesized by physical and chemical methods. Common physical ways include grinding with a mechanical device and using inert gas condensation to cool a liquid[3]. Chemical methods for creating NPs include chemical precipitation, chemical vapor deposition, microemulsions, spray pyrolysis, and thermal spraying. Furthermore, a sonochemistry technique has been developed to produce suspensions of iron NPs stabilized by oleic acid[4]. Metal NPs can now be made by mechanical milling, inert gas condensation, chemical precipitation, spray pyrolysis, and thermal spraying.

NPs are most commonly found in powder form. An aqueous or organic base liquid can be used to dissolve NPs in a powder form to make nanofluids for a variety of purposes.

8.3 STRUCTURE, SURFACE MORPHOLOGY, OPTICAL AND ELECTRICAL PROPERTIES OF COPPER NANOPARTICLES

8.3.1 SURFACE MORPHOLOGIES

Among the many metal particles, copper NPs have gotten a lot of attention since copper is a versatile substance with a wide range of applications in academia and industry. Copper's electrical conductivity is remarkable. In current electronic circuits, copper is used widely because of its low cost[5]. Since copper NPs have such excellent electrical conductivity, catalytic behavior, compatibility, and surface increased Raman scattering activity, scientists have been enthusiastic to incorporate them in future nanodevices[6]. For medical and bioanalytical applications, copper NPs were investigated as potential nanoprobes. Copper NPs and nanowires were made using a variety of techniques[7]. As an alternative catalyst, copper NPs can be employed in a variety of catalytic processes, including selective hydrogenation and methanol synthesis reactions, making them appropriate for use in the field of catalysis[8, 9]. Analysis of the manufactured copper NPs was carried out by means of an array of techniques, including powder-X-ray diffraction analysis and scanning electron microscopy and transmission electron microscopy, as well as an EDX spectrum, DLS, UV-analysis, and dielectric investigations.

8.3.2 X-RAY DIFFRACTION ANALYSIS

The X-ray powder diffraction (XRD) technique is a useful tool for identifying and evaluating the structural characteristics of crystalline phases in materials (strain state, grain size, epitaxy, phase composition, preferred orientation, and defect structure). Utilizing a powder X-ray diffractometer (Schimadzu model: XRD 6000 using CuK) with a diffraction angle spanning from 20° to 80°, the XRD pattern of Cu NPs was recorded. The peaks are exceptionally sharp due to copper's nanocrystalline structure. There are no impurity peaks in the X-ray diffraction pattern. The crystallite size was determined using Scherrer's formula from the broadenings of the respective X-ray spectrum peaks. The average nanocrystalline size was calculated using the Scherrer formula (D).

8.3.3 SCANNING ELECTRON MICROSCOPY (SEM) AND ENERGY-DISPERSIVE X-RAY (EDX) ANALYSIS

The scanning electron microscope (SEM) is one of the most commonly used pieces of equipment for analyzing nanomaterials and nanostructures. Electron-sample interactions generate signals that convey information about the sample's surface morphology (texture) and chemical composition. Cu NPs were spherical in shape and had an average particle size of about 50 nm.

8.3.4 TRANSMISSION ELECTRON MICROSCOPY (TEM)

Transmission electron microscopy (TEM) is frequently used for identifying chemical composition, visualizing the size of small NPs, and exposing phase/crystallographic orientation information through a diffraction pattern. A transmission electron microscope (TEM) image was captured using a Hitachi H-800 TEM with a 100kV accelerating voltage. The scanning transmission electron microscopy (TEM) technique is widely used to image and analyze NPs in order to determine their shape, size, and morphology. In addition to the individual particles, there are certain aggregates. The size of the particles is estimated to be between 30 and 50 nm.

8.3.5 DYNAMIC LIGHT SCATTERING (DLS) STUDIES

DLS is an excellent method for determining the size of NPs in a solution. The aggregation state of the particles can be evaluated by comparing DLS sizing data with transmission electron microscope images. Copper NPs had a particle size of 30 to 50 nm, according to the dynamic light scattering experiment, which was validated by TEM examination.

8.3.6 OPTICAL STUDIES

The process through which the energy of a photon is taken up by matter, most often the electrons of an atom, is known as absorption of electromagnetic radiation. When high-energy light is absorbed by atoms or molecules, creating electronic excitation, UV-visible spectroscopy is used. The synthesized copper NPs have an absorbance peak of roughly 570 nm. The absorption of copper NPs is responsible for this peak. The relationship between the optical absorption coefficient and photon energy is useful in determining band structure and electron transition type.

8.3.7 DIELECTRIC STUDIES

The dielectric analysis is a valuable method for determining the electrical characteristics of materials at different frequencies. At frequencies ranges from 50Hz to 5MHz, the dielectric characteristics of copper NPs were examined using an HIOKI 3532–50 LCR HITESTER.

The dielectric constant and dielectric loss both diminish as frequency increases at room temperature. At low frequencies, the significant dielectric constant demonstrates the presence of space charge polarization at the grain boundary interface[10, 11]. Electronic polarization has a minimal impact on the dielectric constant. The purity of the NPs determines the contribution of the space charge to polarizability[12]. The dielectric loss diminishes with increasing frequency. The curve indicates that, like the dielectric constant, dielectric loss is greatly influenced by the frequency of the applied field.

8.4   BARRIERS AND CHALLENGES TO THE COMMERCIAL PRODUCTION OF NANOFLUIDS

Nanofluids have been developed in modest numbers, although they exhibit a wide range of properties. Commercial applications will benefit from larger-scale production of nanofluids with high dispersion at a reasonable cost.

8.4.1   TWO-STEP PROCESS

In a two-step process NPs may now be produced affordably in large quantities using the inert gas condensation process, which is a benefit of the two-step technique for commercial nanofluid production[8]. It's possible to generate nanofluids in two steps using bulk-produced nanopowders, if the agglomeration issue can be resolved. In order to obtain total dispersion of NPs, two-step synthesis of nanofluids is required. Because of the strong van der Waals forces between the NPs, this agglomeration quickly settles out of the liquid. The usage of oxide NPs, which necessitate larger volume concentrations to achieve the same heat transfer increase in nanofluids as metal particles, exacerbates this barrier. A few surface-treated NPs disperse efficiently in base fluids and have good thermal properties. The agglomeration problem becomes more pronounced at high volume concentrations. NPs that have been coated with a surface treatment can diffuse easily in base fluids and exhibit good thermal conductivity. There are currently commercially available liquid suspensions of small particles for a number of different fluids. Ceramic suspensions are widely accessible. A two-step procedure is used to make nanofluids in the laboratory, and these fluids have the same agglomeration concerns as those created in the lab.

8.4.2   ONE-STEP PROCESS

It is better to use a one-step procedure than a two-step process to avoid particles from oxidizing in nanofluids containing high-conductivity metals like copper. Through the use of this technology, NPs are created and disseminated within fluids in a single step. Non-agglomerating copper NPs in ethylene glycol were synthesized by direct evaporation in a one-step process[9]. Another one-step physical process, submerged arc NP synthesis, has been utilized to generate nanofluids containing diverse NPs such as TiO_2, CuO, and Cu[10–12]. The nanofluid is formed by arc sparking the solid material from an electrode and then condensing it into a liquid in a vacuum chamber.

It is unlikely that one-step physical procedures will become the dominant way of commercial nanofluid synthesis, even though they have produced small amounts for study. Two considerations make it incredibly difficult to scale. A significant impact on progress is had by applying vacuum technologies to slow down the production of nanomaterials and fluids. These one-step physical techniques also have a significant cost in terms of manufacturing nanofluids.

It is claimed that the reduction of a copper salt by sodium hypophosphite generated and disseminated virtually mono distributed copper NPs in ethylene glycol. For protection and stabilization, polyvinylpyrrolidone (PVP) was utilized in the formulation. This technology has the potential to manufacture enormous quantities of nanofluids more quickly than the one-step physical process in the future with additional improvement. However, this technology has a significant drawback in that it can only create much lower concentrations of NPs and nanofluids than the two-step method.

When producing nanofluids in a batch mode, both physical and chemical one-step procedures have limited control over a number of essential factors such as those that affect NP size. If a one-step chemical reaction could be repeated constantly, its commercial viability would increase. An ultra thin organic layer (2–10 nm) coated with copper NPs and chemically compatible with the host fluid was recently created to achieve suspension stabilization using a semi-continuous technique for the one-step synthesis of copper nanofluids based on ethylene glycol[13].

8.4.3 OTHER PROCESSES

It is possible to produce NPs with precise geometries and densities, as well as porosities and surface chemistries using processes such as electrolytic metal deposition and layer-by-layer assembly. Another technology, chemical vapor condensation, appears to have advantages in terms of particle size control, simplicity of scalability, and the capacity to construct unique core-shell nanostructures[14]. Another option is to adjust the form and size of NPs at room temperature[15]. The synthesis technique has an impact on the structural properties of NPs, such as the mean particle size, particle size distribution, and shape, and there is space for improvement. Regardless of the method employed to make nanofluids, the properties of the resulting particles in suspension are challenging to assess.

REFERENCES

[1] Cheng, L., Bandarra Filho, E. P. & Thome, J. R., 'Nanofluid two-phase flow and thermal physics: A new research frontier of nanotechnology and its challenges', *J. Nanosci. Nanotechnol.*, vol. 8, no. 7, pp. 3315–3332, 2008.

[2] Cheng, L. & Liu, L., 'Boiling and two phase flow phenomena of refrigerant-based nanofluids: Fundamentals, applications and challenges', *Int. J. Refrig.*, vol. 36, no. 2, pp. 421–446, 2013.

[3] Thome, J. R., 'The new frontier in heat transfer: microscale and nanoscale technologies', *Heat Transfer Eng.*, vol. 27, no. 9, pp. 1–3, 2006.

[4] Cheng, L. & Xia, G., 'Fundamental issues, mechanisms and models of flow boiling heat transfer in microscale channels', *Int. J. Heat Mass Transfer*, vol. 108, Part A, pp. 97–127, 2017.

[5] Thome, J. R., 'Boiling in microchannels: A review of experiment and theory', *Int. J. Heat Fluid Flow*, vol. 25, no. 2, pp. 128–139, 2004. 80.

[6] Cheng, L., 'Fundamental issues of critical heat flux phenomena during flow boiling in microscalechannels and nucleate pool boiling in confined spaces', *Heat Transfer Eng.*, vol. 34, no. 13, pp. 1011–1043, 2013.

[7] Cheng, L. & Mewes, D., 'Review of two-phase flow and flow boiling of mixtures in small and mini channels', *Int. J. Multiphase Flow*, vol. 32, no. 2, pp. 183–207, 2006.

[8] Cheng, L. & Thome, J. R., 'Cooling of microprocessors using flow boiling of CO_2 in a microevaporator: Preliminary analysis and performance comparison', *Appl. Therm. Eng.*, vol. 29, no. 11–12, pp. 2426–2432, 2009.

[9] Thome, J. R., *Enhanced Boiling Heat Transfer*. New York: Hemisphere Publ. Corp., 1990.

[10] Xia, G., Du, M., Cheng, L. & Wang, W., 'Experimental study on the nucleate boiling heat transfer characteristics of a multi-wall carbon nanotubes water based nanofluid in a confined space', *Int. J. Heat Mass Transfer*, vol. 113, pp. 59–69, 2017.

[11] Cheng, L., 'Flow boiling heat transfer and critical heat flux phenomena of nanofluids in microscale channels', *Int. J. Microscale Nanoscale Therm. Fluid Transp. Phenom*, vol. 5, no. 3–4, pp. 201–214, 2014.

[12] L. Cheng, D. M. & Luke, A., 'Boiling phenomena with surfactants and polymeric additives: A state-of-the-art review', *Int. J. Heat Mass Transfer*, vol. 50, no. 13–14, pp. 2744–2771, 2007. 81.

[13] Xuan, Y. & Li, Q., 'Heat transfer enhancement of nanofluids', *Int. J. Heat Fluid Flow*, vol. 21, no. 1, pp. 58–64, 2000.

[14] Das, S. K., Putra, N., Thiesen, P. & Roetzel, W., 'Temperature dependence of thermal conductivity enhancement for nanofluids', *ASME J. Heat Transfer*, vol. 125, no. 4, pp. 567–574, 2003.

[15] Lee, S., Choi, S. U. S., Li, S. & Eastman, J. A., 'Measuring thermal conductivity of fluids containing oxide nanoparticles', *ASME J. Heat Transfer*, vol. 121, no. 2, pp. 280–289, 1999.

9 Thermal Conductivity of Nanofluids

CONTENTS

9.1 INTRODUCTION

Thermal conductivity enhancement of nanofluids is consistent for a longer period to which the surfactants are added to them, in order to diminish the aggregation of the nanoparticles. In this procedure, a small amount of acceptable surfactant is added to the base fluid and swirled constantly for a few hours, approximately one-tenth of the mass of NPs. Surfactant-based nanofluids produce a stable suspension in the host liquid with uniform particle dispersion. The NPs remain floating in the container for a long time without dropping to the bottom.

The NPs are mixed in the base fluid of water-ethylene glycol mixture after determining the amount of NPs needed to prepare CuO nanofluid for a specific volume concentration. Surfactants and acid are not used in this investigation because surfactants alter the thermophysical characteristics of nanofluids. Because corrosion happens after a few days of continued use of such nanofluids in practical applications, the addition of acid to the tube material may cause harm to the tube material. For testing the temperature-dependent thermal conductivity and viscosity of all nanofluid concentrations considered, copper oxide nanofluids with volume concentrations of 0.025%, 0.1%, 0.4%, 0.8%, and 1.25% are created. When nanoparticles are suspended in a base fluid, they typically clump together. When NPs are suspended in a base fluid, they agglomerate.

The isentropic, Newtonian behavior of the CuO, Al_2O_3 nanofluids generated is assumed, with thermophysical parameters that are uniform and consistent throughout time across the fluid sample[1].

9.2   EFFECTS OF NANOPARTICLES ON THE cP

Heat transfer fluids must meet specific characteristics in addition to being affordable. They should have thermal stability at high temperatures. These materials should have the lowest feasible melting point. At high temperatures, the same can be said of their vapor pressure (700°C). Corrosivity, viscosity, thermal conductivity, and toxicity are just a few of the key characteristics to consider.

The heat transfer fluid's thermodynamic property of specific heat capacity should be as high as feasible in J/g K. The molten salts or salt combinations indicated earlier have a wide temperature range of useful use, which outweighs their low specific heat capacity of 0.75 to 2 J/kg K. The specific heat capacity at constant pressure and the specific heat capacity of the base salt mixture is greatly increased by particles with a size distribution peaking about 10 to 100 nm. Nanofluids are a larger, more comprehensive category of fluids that includes salt solutions doped with NPs.

To summarize, every shape of NP utilized in salt combinations enhances their specific heat capacity. Particle mass concentration should not, however, exceed 1% of the overall weight. The preparation procedures' specifics may differ (e.g. the temperatures during drying of the samples). Regardless of these differences, average cP increases range from 10% to 30%. Water doped with NPs has so far resulted in a reduction in heat capacity[2–4].

9.2.1   The Rationale behind Nanofluids

According to a study of thermal characteristics, all liquid coolants utilized as heat transfer fluids today have extremely low thermal conductivity. Engine coolants, lubricants, and organic coolants, for example, conduct heat three orders of magnitude slower than copper. Any efforts to increase heat transfer by creating turbulence, increasing area, and so on will be limited by the fluid's inherent thermal conductivity restriction. The thermal conductivity of slurries can be accurately predicted using these methods. However, all of this research focused on the suspension of micro- to macro-sized particles, which has the following fundamental limitations[5].

1. The particles immediately settle to the surface, generating a layer that reduces the fluid's heat transfer capacity.
2. Sedimentation diminishes as the fluid circulation rate rises, but erosion of heat transfer devices, pipelines, and other structures increases.
3. Large particles obstruct flow channels, especially if the cooling channels are narrow.
4. The fluid's pressure drop increases significantly.
5. Finally, particle concentration-based conductivity enhancement is achieved (i.e. the bigger the particle volume fraction, the stronger the enhancement— and the greater the issues, as seen in 1–4).

Consequently, the approach of suspending particles in liquid was well-known but rejected for heat transfer applications. Although nanofluids have made this scientists reconsider this choice, it is still a viable option. Materials science has advanced to the point where nanometer-sized particles having mechanical, thermal, electrical, and optical properties can now be created. Due to recent advances in nanofluid technology[6] and heat transfer research on microchannel flow, a new method for studying nanoparticle suspensions is now feasible. Despite the fact that all fluids are "nano" because of their molecular chains, this classification has been widely recognized and popular in the scientific world. Nanofluids, such as DNA, RNA, proteins, and fluids confined in nanopores, have all been referred to by biologists under the term "nanofluid"[7–9]. NPs are desirable because of their huge surface area, low particle momentum, and immense mobility. The conductivity of MWCNTs, which is 20,000 times greater than that of engine oil at room temperature, can be improved upon by starting with copper[10].

When the nanofluid particles are appropriately dispersed, these features are expected to provide the following advantages:

1. **Improved heat conduction**. Increased heat transfer is made possible by the nanoparticles' enormous surface area. At the surface of particles smaller than 20 nm, 20% of their atoms are immediately available for thermal interaction. It is also possible to increase heat transfer because of the small size of the particles; this is due to micro-convection[11].

2. **Constancy**. It is less likely that sedimentation will occur due to the smaller size of the particles. There is less sedimentation since the particles are smaller and therefore lighter.

3. **Non-clogging microchannel cooling**. Nanofluids will not only be a better heat transfer medium, but they will also be suited for microchannel applications with high heat loads. More heat can be transferred in a larger area by combining nanofluid and microchannel systems. Because microchannels are clogged with meso- or microparticles, this cannot be achieved. Unlike microchannels, nanoparticles have only a few hundred or thousands of atoms each.

4. **Less chance of erosion**. The momentum that NPs may transmit to a solid wall is even less than the particles themselves, at sub-atomic sizes. Components, such as heat exchangers, pipes, and pumps, are less likely to be eroded due to the lower momentum.

5. **Reduced pumping power**. In order to double the heat transfer efficiency of traditional fluid, an increase in pumping power of ten times is often required. The same apparatus's heat transfer may be proved to be three times as efficient when the conductivity is increased three times as much. As long as fluid viscosity doesn't significantly rise, pumping power will only need to be mildly increased. Consequently, pumping power can be saved if considerable thermal conductivity increases can be accomplished with modest volume fractions of particles.

Nature can often be more interesting than imagination at times. Prior to the discovery of radioactivity, Henry Becquerel believed that uranium ore absorbed sunlight before re-radiating it. Similar to the first test with nanofluids, the results were unexpectedly encouraging. The following are the four most significant outcomes:

1. **Increased thermal conductivity that is not expected**. The most notice-able property of nanofluids was an extraordinary rise in thermal conductiv-ity that exceeded expectations and was significantly higher than any theory could anticipate.
2. **Constancy**. A stabilizing ingredient has been found to keep nanofluids stable for months[12, 13].
3. **Typical Newtonian behavior plus a small amount of concentration**. With a very low particle concentration, a considerable improvement in con-ductivity was achieved while preserving the fluid's Newtonian properties. Because the rise in viscosity was so little, the pressure drop was also small.
4. **The size of the particles**. For microslurries, particle concentration was shown to be important, but particle size was found to be even more impor-tant in increasing conductivity. There has been a rise in enhancement with decreasing particle size, on the whole.

Assuming that nanofluids will play an important role in building next-generation cooling solutions, the development of nanofluids begins. Preliminary results show that this is an interdisciplinary field with contributions from many different scientific disciplines, including chemistry, mechanical and chemical engineering, physics, and material science. It is therefore worthwhile to examine the details of not only appli-cations, but also the processes of synthesis and characterization.

9.2.2 Synthesis and Preparation of Nanoparticles

The thermal conductivity of nanofluids is much higher than that of regular suspen-sions. In nanofluids, the thermal conductivity gain over the base fluid is typically many times greater than in micrometer-sized suspensions. Fluids such as ethylene glycol, transformer oil, and toluene have so far been employed. Carbon nanotubes, ceramic particles, and pure metallic nanoparticles are the most often employed NPs (CNTs). Different nanofluids can be created by combining the particles and fluids in a variety of ways[14].

9.2.3 Ceramic Nanofluids

Water and ethylene glycol NPs were tested for conductivity in the field's first sub-stantial publication[12]. Conductivity was measured using the standard transient hot-wire (THW) method. It was obvious from the results that the thermal conductivity of Al_2O_3 and CuO nanofluids had been significantly improved. It is employed in volu-metric fractions ranging from 1% to 5%. Ethylene glycol was used as the basis fluid, which resulted in a maximum improvement. An improvement of 20% was recorded at a volume percentage of 4% CuO. When water was utilized as the base fluid, the augmentation was lower but still considerable, with a 12% increase at 3.5% CuO and a 10% increase at 4% Al_2O_3[12]. Although Maxwell's suspension model was revised in 1962 to include particle shape, these results outperformed Maxwell's model for suspensions. As a weighted average of solid and liquid conductivity, these models predict effective thermal conductivity.

9.2.4 METALLIC NANOFLUIDS

The development of metallic particle-based nanofluids, despite the fact that ceramic nanofluids had already proved the promise of nanofluids, was an important step forward. This was the first time that copper-based transformer oil nanofluids had been used to test a new theory. However, despite the use of significantly larger (100 nm) particles, the observed increase in volume fraction was 55%.

This achievement was made possible by the ANL group, which found that a 0.3% concentration of copper NPs in ethylene glycol increased conductivity by 40.0%[15]. There was a clear demonstration of the effect of particle size and the promise of nanofluids with smaller particles. The nanofluids were stabilized using thioglycolic acid. While the conductivity of toluene-gold nanofluid improved by 3–7% at a volume fraction of only 0.005–0.011%, the rise for water-gold nanofluid was only 3.2% for a volume fraction of 0.0013–0.0026% at ambient temperature.

Particles with a diameter of 10–20 nm were the greatest contributor to the improvement. The improvement was greater in water-based nanofluids due to the usage of bare particles and lower in toluene-based nanofluids due to the nanoparticles' protection from agglomeration by a layer of thiolate coating. The conductivity of water silver nanofluids was also found to be lower than that of conventional water silver nanofluids. In spite of its better conductivity, silver gave less enhancement due to its larger size (60–80 nm), as it was clearly shown. The conductivity or concentration of particles may not matter as much as the particle size. Dispersions of nano-sized -Al_2O_3 in diverse solvents, including deionized water, glycerol, pump oil and ethylene glycol and water mixtures as well as glycerol and water. Higher base fluid thermal conductivity results in lower thermal conductivity ratios.

Ethylene glycols enhance the thermal conductivity of nanofluids containing 10 nm-sized Fe NPs by a factor of 10. Without increasing volume fraction, an 18% improvement is achieved. To prove that particle size has a direct effect on the thermal conductivity of the nanofluid, sonication of the nanofluid was found to have a considerable impact[16].

9.2.5 CARBON AND POLYMER NANOTUBE NANOFLUIDS

Two things can be considered: enhancement and nonlinear behavior. First, carbon nanotubes have very high thermal conductivity (3000 W/mK) and second, the nanotubes have a very high aspect ratio (2000). The aspect ratio of nanotubes will be discussed in this article as it pertains to possible theories of nanofluid thermal conductivity.

An experiment is conducted to determine the thermal conductivity of 15-nm MWCNTs suspended in water, ethylene glycol, and decene, with a 30-meter length[17]. In water and ethylene glycol, the surfactant-free suspensions were covered with oxygen-containing functional groups to prevent settling. Oleylamine served as a surfactant in the decene solution to help keep the particles suspended. Compared to fluids with a higher thermal conductivity, fluids with a smaller volume fraction exhibited greater improvement. Adding 1% volume of CNTs improved decene thermal conductivity by 20%. It was also found to rise linearly with the volume fraction. When suspended in water at 1% volume, MWCNTs with an average diameter and length of 130 nm showed a 34% thermal conductivity while double-walled CNTs had an 8% value. CNTs with diameters of 20–50 nm were found to increase the thermal conductivity of ethylene glycol by 12.4% and synthetic oil by 30% when the volume fraction of CNTs in each medium was increased from 1% to 2%, respectively, using a thermal conductivity measurement technique.

9.3 RESEARCH ON THERMAL CONDUCTIVITIES OF NANOFLUIDS

Table 9.1 shows that the thermal conductivities of solids are several orders of magnitude greater than those of traditional heat transfer fluids. Nanofluids can be considerably improved by suspending NPs in conventional heat transfer fluids. It is predicted that the heat transfer performance will be much improved.

Nanofluids' heat transfer-enhancing properties may be explained by the following factors:

1. The surface area and heat capacity of the fluid are both increased as a result of the NP suspension.
2. The thermal conductivity of the fluid is enhanced by the presence of suspended NPs.
3. Particle, fluid, and flow passage surface collisions and interactions are increased.
4. Turbulence in the fluid is increased.
5. Dispersion of NPs in the fluid, which results in a more even distribution of temperature.

TABLE 9.1
Thermal Conductivities of Various Solids and Liquids at Room Temperature

Material	Form	Thermal Conductivity (W/mK)
Carbon	Nanotubes	1800–6600
	Diamond	2300
	Graphite	110–190
	Fullerenes film	0.4
Metallic solids (pure)	Silver	429
	Copper	401
	Nickel	237
Non-metallic solids	Silicon	148
Metallic liquids	Aluminum	40
	Sodium at 644 K	72.3
Others	Water	0.613
	Ethylene Glycol	0.253
	Engine Oil	0.145
	R134a	0.0811

A nanofluid containing a small amount of thioglycolic acid improved the stability of the metallic particles against settling the most significantly. The heat transfer properties of nanofluids are significantly superior to those of ordinary fluids. Increased particle density raises the nanofluid's thermal conductivity (the ability to conduct heat).

In general, metallic nanofluids have a much better impact on performance than metallic oxide nanofluids. Additionally, nanofluids' thermal conductivity varies with temperature. A number of factors, including particle size, shape, and volume concentration, have an impact on the thermal conductivity of nanofluids[18]. In a Fe nanofluid, the thermal conductivity increases up to 18%, when the particle volume fraction rises to 0.55 vol%. The suspension of highly thermally conductive materials is not always effective, as was discovered when comparing Fe nanofluids to Cu nanofluids.

Nanofluid CNT NPs have been extensively studied because of their outstanding mechanical and electrical capabilities, which can be attributed to their distinctive structural features. Compared to a liquid refrigerant, CNTs have a very high thermal conductivity of up to 6600 W/mK[19–22]. The thermal conductivity and heat transfer properties of nanofluids can be improved by using CNTs[23–26]. Thermal conductivity of a SWCNT-polymer epoxy composite increased by 70% at 40 K and increased by 125% at room temperature with a 1% nanotube loading[23].

Carbon nanotubes have an incredibly high aspect ratio. The viscosity measurements suggest that CNTs from a densely entangled fiber network are not very mobile, and hence their influence on thermal transport in fluid suspensions is likely to be similar to that of polymer composites. The thermal conductivity of SWCNTs grew linearly with temperature from 7 K to 25 K, increased in slope from 24 K to 40 K, and then rose monotonically with temperature until it reached temperatures above room temperature. A temperature-dependent thermal conductivity of 6600 W/mK is achieved by the carbon nanotubes[21]. At ambient temperature, the thermal conductivity of individual multiwalled nanotubes exceeded 3000 W/mK[22, 23].

Thermal conductivity of nanotube suspensions is not linearly related to nanotube volume percent as predicted theoretically (inset); this is a major exodus from expectations. Because of the increase in concentration, the thermal conductivity increases considerably. In pure water and ethylene glycol, the CNT suspensions had 10–20% better effective thermal conductivities[24–26]. Because there is no theory for the thermal conductivities of nanofluids, existing thermal conductivity models for conventional solid/liquid systems have been used to determine the effective conductivities of nanofluids.

Thermal conductivity may be improved by the Brownian motion of the NPs in these solutions[15]. For example, a theoretical model was created to explain the essential function of NP dynamics in nanofluids. Besides temperature and concentration, the model predicts a major size-dependent effect.

Nanofluid heat transfer may be enhanced by one or more of the following methods:

1. Transport of ballistic phonons inside NPs.
2. Liquid molecule layering at the interfacial surface.
3. Brownian motion of NPs, and NP clustering.

It's difficult to develop a theoretical thermal conductivity model based on these principles because the physical mechanics involve many unknowns. In addition to measuring the thermal conductivity of the NP material and the size, shape, and volume fraction of the NPs, measurements should also be taken of the thermal conductivity of the base fluid and the thermal conductivity of the NPs. Brownian motion also has a direct relationship with other nanofluid properties like base fluid diffusion velocity and nanofluid particle aggregation.

REFERENCES

[1] Das, S. K., Choi, S. U. S.& Patel, H., 'Heat Transfer in nanofluids-a review', *Heat Transf. Eng.*, vol. 27, no. 10, pp. 3–19, 2006.

[2] Fang, X., Chen, Y., Zhang, H., Chen, W., Dong, A. & Wang, R., 'Heat transfer and critical heat flux of nanofluid boiling: A comprehensive review', *Renew. Sustain. Energy Rev.*, vol. 62, pp. 924–940, 2016.

[3] Cheng, L., 'Nanofluid heat transfer technologies', *Recent Patents Eng.*, vol. 3, no. 1, pp. 1–7, 2009.

[4] Cheng, L., 'Fundamental issues of critical heat flux phenomena during flow boiling in microscalechannels and nucleate pool boiling in confined spaces', *Heat Transfer Eng.*, vol. 34, no. 13, pp. 1011–1043, 2013.

[5] Cheng, L. & Mewes, D., 'Review of two-phase flow and flow boiling of mixtures in small and mini channels', *Int. J. Multiphase Flow*, vol. 32, no. 2, pp. 183–207, 2006.

[6] Thome, J. R., 'The new frontier in heat transfer: Microscale and nanoscale technologies', *Heat Transfer Eng.*, vol. 27, no. 9, pp. 1–3, 2006.

[7] Cheng, L. & Thome, J. R., 'Cooling of microprocessors using flow boiling of CO2 in a microevaporator: Preliminary analysis and performance comparison', *Appl. Therm. Eng.*, vol. 29, no. 11–12, pp. 2426–2432, 2009.

[8] Thome, J. R., *Enhanced Boiling Heat Transfer*. New York: Hemisphere Publ. Corp., 1990.

[9] Xia, G., Dum, M., Cheng, L. & Wang, W., 'Experimental study on the nucleate boiling heat transfer characteristics of a multi-wall carbon nanotubes water based nanofluid in a confined space', *Int. J. Heat Mass Transfer*, vol. 113, pp. 59–69, 2017.

[10] Cheng, L., 'Flow boiling heat transfer and critical heat flux phenomena of nanofluids in microscale channels', *Int. J. Microscale Nanoscale Therm. Fluid Transp. Phenom*, vol. 5, no. 3–4, pp. 201–214, 2014.

[11] Cheng, L., Mewes, D. & Luke, A., 'Boiling phenomena with surfactants and polymeric additives: A state-of-the-art review', *Int. J. Heat Mass Transfer*, vol. 50, no. 13–14, pp. 2744–2771, 2007. 81.

[12] Eastman, J. A., Choi, S. U. S., Li, S., Yu, W. & Thompson, L. J., 'Anomalously increased effective thermal conductivities of ethylene glycol-based nanofluids containing copper nanoparticles', *Appl. Phys. Lett.*, vol. 78, no. 6, pp. 718–720, 2001.

[13] Xuan, Y. & Li, Q., 'Heat transfer enhancement of nanofluids', *Int. J. Heat Fluid Flow*, vol. 21, no. 1, pp. 58–64, 2000.

[14] Das, S. K., Putra, N., Thiesen, P. & Roetzel, W., 'Temperature dependence of thermal conductivity enhancement for nanofluids', *ASME J. Heat Transfer*, vol. 125, no. 4, pp. 567–574, 2003.

[15] Jang, S. P.& Choi, S. U. S., 'Role of Brownian motion in the enhanced thermal conductivity of nanofluids', *Appl. Phys. Lett.*, vol. 84, no. 21, pp. 219–246, 2004.

[16] Kim, J. H., Kim, K. H.& You, S. M., 'Pool boiling heat transfer in saturated nanofluids', *ASME Int. Mech. Eng. Congr. & Exp.*, Anaheim, Californian, USA, Heat Transfer, vol. 2, pp. 621–628, Nov. 13–19, 2004.

[17] Assael, M. J., Metaxa, I. N., Arvanitidis, J., Christofilos, D. & Lioutas, C., 'Thermal conductivity enhancement in aqueous suspensions of carbon multi-walled and double-walled nanotubes in the presence of two different dispersants', *Int. J. Thermophys.*, vol. 26, no. 3, pp. 647–664, 2005.

[18] Murshed, S. M. S., Leong, K. C. & Yang, C., 'Investigations of thermal conductivity and viscosity of nanofluids', *Int. J. Therm. Sci.*, vol. 47, no. 5, pp. 560–568, 2008.

[19] Biercuk, M. J., Llaguno, M. C., Radosavljevic, M., Hyun, J. K.& Johnson, A. T., 'Carbon nanotube composites for thermal management', *Appl. Phys. Lett.*, vol. 80, no. 15, pp. 2767–2769, 2002.

[20] Hone, J., Whitney, M. & Zettl, A., 'Thermal conductivity of single-walled carbon nanotubes', *Synth. Met.*, vol. 103, no. 1–3, pp. 2498–2499, 1999.

[21] Berber, S., Kwon, Y. K.& Tomanek, D., 'Unusually high thermal conductivity of carbon nanotubes', *Phys. Rev. Lett.*, vol. 84, no. 20, pp. 4613–4616, 2000.

[22] Kim, P., Shi, L., Majumdar, A. & Mceuen, P. L., 'Thermal transport measurements of individual multiwalled nanotubes', *Phys. Rev. Lett.*, vol. 87, no. 21, pp. 215502-1–215502-4, 2001.

[23] Choi, S. U. S., Zhang, Z. G., Yu, W., Lockwood, F. E.& Grulke, E. A., 'Anomalous thermal conductivity enhancement in nanotube suspensions', *Appl. Phys. Lett.*, vol. 79, no. 14, pp. 2252–2254, 2001.

[24] Xie, H., Lee, H., Youn, W. & Choi, M., 'Nanofluids containing multiwalled carbon nanotubes and their enhanced thermal conductivities', *J. Appl. Phys.*, vol. 94, no. 8, pp. 4967–4971, 2003.

[25] Liu, M. S., Lin, M. C. C., Huang, I. T.& Wang, C. C., 'Enhancement of thermal conductivity with carbon nanotube for nanofluids', *Int. Comm. Heat Mass Transfer*, vol. 32, no. 9, pp. 1202–1210, 2005.

[26] Wen, D. S.& Ding, Y. L., 'Effective thermal conductivity of aqueous suspensions of carbon nanotubes nanofluids', *J. Thermophys. Heat Transfer*, vol. 18, no. 4, pp. 481–485, 2004.

10 Thermophysical Properties of Nanofluids

CONTENTS

10.1 INTRODUCTION

The rheological properties of CuO-water nanofluids are studied at temperatures ranging from 278 to 323 K with volumetric contents of 5% to 15%[1, 2]. There was no time-dependent shear thinned or pseudo-plastic non-Newtonian fluid properties shown by these nanofluids. However, the model may not be relevant to other nanofluids and situations because of the diverse behaviors of different nanofluids.

There is a point at which nanofluid viscosity changes substantially for a given particle volume concentration. The viscosity increases considerably when a fluid sample is heated above this crucial temperature. A hysteresis phenomenon happens when it cools after being heated over this critical temperature. Hysteresis is an intriguing phenomenon for which no mechanisms have been identified[3].

The viscosity of water-based CNT nanofluids increased with increasing CNT content and lowering temperature. At a pH of 6, the CNT nanofluids' viscosity data are displayed in Figure 10.1. A non-Newtonian nanofluid, the CNT nanofluid shows shear thinning behavior. A nonlinear relationship arises when shear rates are large.

With reference to the theoretical model for the viscosity of solid-fluid mixtures, the standard Einstein equation[5] determines the fluid's effective viscosity:

$$\mu_{mix} = \mu_f \left(1 + \frac{5}{2}\phi\right)$$

where μ_{mix} denotes the viscosity of the mixed fluid,
μ_f the viscosity of the ambient fluid,
and Φ is the volume fraction of spheres in suspension.

Since these models solely connect viscosity to volume concentration, experimentally reported nanofluid viscosities depart from the classical model. The temperature dependence and particle aggregation are not included in this model.

DOI: 10.1201/9781003163633-10

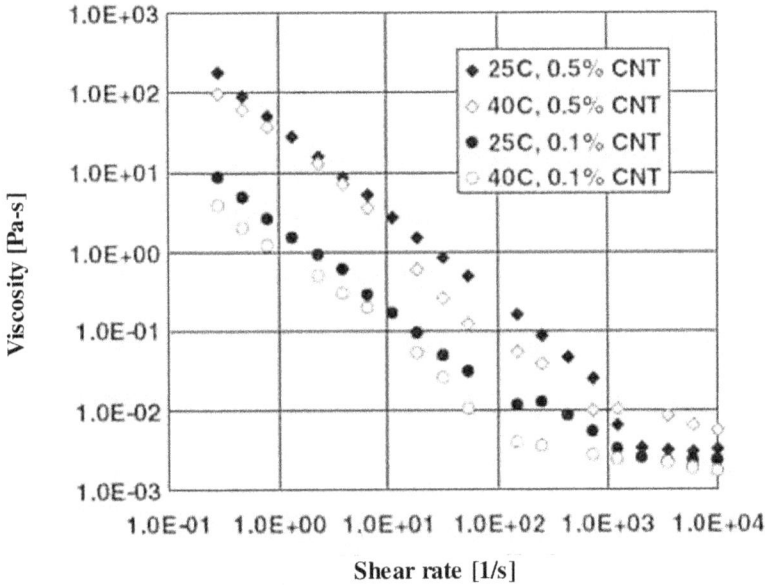

FIGURE 10.1   Viscosity of CNT nanofluids (pH = 6)[4].

Nanofluid viscosity remains a mystery since the fundamental physical principles are unknown. Because of these difficulties, developing nanofluid models and prediction tools can be challenging.

Several models and methods for nanofluid viscosity prediction have been established, but there are considerable variations between measured and calculated values for various nanofluids. This is mainly because of the governing parameters that can have a substantial impact on the fluid behavior, such as NP materials, NP size and shape, concentration, pH, and temperatures, among others. For the viscosities of nanofluids, there is however no systematic theory or generalized model. In terms of nanofluid viscosity, there are numerous influences that can have a significant impact. Nanofluids are frequently made using surfactants as a stabilizing agent.

10.2   RESEARCH ON THE SPECIFIC HEATS OF NANOFLUIDS

The specific heat of a nanofluid is determined by the specific heat of the base fluid and the NPs, the volume concentration of NPs, and the temperature. Differential scanning calorimeters are utilized to assess the specific heat capacity of exfoliated graphite NPs suspended in polyalphaole fins at concentrations of 0.6 and 0.3% mass, respectively. It was found that the specific heat increased with increasing temperature. Nanofluid heat transfer and heat transfer mechanisms will benefit from this research. In addition, physical mechanisms for the augmentation of specific heat are urgently required to explain the experimental outcomes.

There were three nanofluids with Al_2O_3, ZnO, and SiO_2 NPs that were tested for their specific heat[6]. The first two were dissolved in ethylene glycol/water (60:40 mass ratio) and the third in deionized water. Reliable models for forecasting the specific heat of nanofluids need to be developed as soon as possible. The specific heat of nanofluids has been modeled using a variety of approaches. The thermal characteristics of a solution containing NPs are different from those of the dispersed and fluid phases when they are combined. Two simple analytical methods based on classical mixing theory have been suggested for determining the specific heat of nanofluids.

The first model assumes a linear relationship, as follows:

$$C_{pnf} = \varphi C_{pnp} + (1-\varphi)C_{pbf}$$

where c = specific heat, subscripts nf, np, and bf = nanofluid, NP, and base fluid, respectively.

Many investigations on heat transfer behavior with nanofluids have utilized a model like this one to estimate nanofluid specific heat, but this model has been found to be insufficient. A local thermal equilibrium is taken into account when computing the specific heat of nanofluids using the following equation[7]:

$$\left(\rho C_p\right)_{nf} = \varphi\left(\rho C_p\right)_{np} + (1-\varphi)\left(\rho C_p\right)_{bf}$$

where the optimal mixture rule is used to predict the density of nanofluids:

$$\rho_{nf} = \varphi\rho_{np} + (1-\varphi)\rho_{bf}$$

In spite of its application in studying nanofluid thermal conductivity, diffusivity, and heat transfer, it has yet to be proven because of several fundamental characteristics that influence the specific heat of nanofluids.

Calculating the specific heat of nanofluids with a more accurate model, this model makes the following assumption about particle-fluid thermal equilibrium:

$$C_{pnf} = \varphi(C_{pnp}) + (1-\phi)C_{pbf}$$

The specific heat capacity of NPs increases with decreasing particle size. Systematic experiments on the specific heat of nanofluids should be performed in order to clarify the mechanisms of specific heat augmentation. In spite of the fact that several models are used to analyze particular heats, the specific temperatures of nanofluids are not well-established. For the prediction of nanofluid temperatures, a generalized model should be developed based on a well-documented database that covers the majority of nanofluids and a wide range of conditions[8–10].

REFERENCES

[1] Kulkarni, D. P., Das, D. K. & Patil, S. L., 'Effect of temperature on rheological properties of copper oxide nanoparticles dispersed in propylene glycol and water mixture', *J. Nanosci. Nanotechnol.*, vol. 7, no. 7, pp. 2318–2322, 2007.

[2] Kulkarni, D. P., Das, D. K. & Chukwu, G. A., 'Temperature dependent rheological property of copper oxide nanoparticles (nanofluid)', *J. Nanosci. Nanotechnol.*, vol. 6, no. 4, pp. 1150–1154, 2006. 84.

[3] Nguyen, C. T., Desgranges, F., Galanis, N., Roy, G., Maré, T., Boucher, S. & Minsta, H. A., 'Viscosity data for Al₂O₃-water nanofluid-hysteresis: Is heat transfer enhancement using nanofluids reliable?', *Int. J. Thermal Sci.*, vol. 47, no. 2, pp. 103–111, 2008.

[4] Ding, Y., Alias, H., Wen, D. & Williams, R. A., 'Heat transfer of aqueous suspensions of carbon nanotubes', *Int. J. Heat Mass Transfer*, vol. 49, no. 1–2, pp. 240–250, 2006. 83.

[5] Einstein, A., 'Eine neue bestimmung der moleküldimensionen', *Ann. Physik.*, vol. 324, no. 2, pp. 289–306, 1906.

[6] Vajjha, R. S. & Das, D. K., 'Specific heat measurement of three nanofluids and development of new correlations', *ASME J. Heat Transfer*, vol. 131, no. 7, pp. 071601-1–071601-7, 2009.

[7] Xuan, Y. & Roetzel, W., 'Conceptions for heat transfer correlation of nanofluids', *Int. J. Heat Mass Transfer*, vol. 43, no. 19, pp. 3701–3707, 2000.

[8] Buobgiorno, J., 'Convective transport in nanofluid', *ASME J. Heat Transfer*, vol. 128, no. 3, pp. 240–250, 2005. 85.

[9] Xue, H. S., Fan, J. R., Hu, Y. C., Hong, R. H. & Cen, K. F., 'The interface effect of carbon nanotube suspension on thermal performance of a two-phase closed thermosyphon', *J. Appl. Phys.*, vol. 100, no. 10, pp. 104909-1-104909-5, 2006.

[10] Tanvir, S. & Qiao, L., 'Surface tension of Nanofluid-type fuels containing suspended nanomaterials', *Nanoscale Res. Lett.*, vol. 7, paper 226, pp. 1–10, 2012.

11 Stability of Nanofluid

CONTENTS

11.1 INTRODUCTION

In the case of nanofluids, long-term stability is a critical issue. In nanofluids, van der Waals forces of attraction cause particles to collect and settle due to the large density difference between them and the basefluid. Also, the properties of the suspended particles and the base fluid are critical to nanofluid stability. The sedimentation velocity (V) is determined by Stokes law[1] as

$$V = \frac{a^2 \left(\rho_{np} - \rho_{bf} \right) g}{g * \eta}$$

where "a"—radius of the dispersed particle,
ρ_{nf}—density of the nanofluid,
ρ_{np}—density of the NP,
η—dynamic viscosity of the nanofluid,
G—gravitational constant.

As the NP size drops, the density gap between the NP and the base fluid narrows, or the base fluid viscosity increases, the sedimentation velocity reduces.

11.2 NANOFLUID STABILITY EVALUATION METHODS

11.2.1 LIGHT SCATTERING TECHNIQUES

Characterizing dispersions with the light scattering technique is common practice. The stability of the nanofluid is impacted by the aggregation and sedimentation generated by particle attractions. The aggregation of particles can also be tracked by measuring the change in particle size and particle size distribution over time. The

DOI: 10.1201/9781003163633-11

SEM or TEM is used to observe NPs after drying; therefore it may not provide an accurate picture of particle shape in fluids. This approach is used to measure tens to hundreds of particles in vast numbers of particles. However, the Dynamic Light Scattering (DLS) approach directly measures particle size over a better sample volume. In DLS, a laser beam is focused on a sample and the dispersed laser light is collected[2-5].

Particles in suspension in liquids are subject to random thermal motion due to Brownian motion, and the velocity at which they move depends on the particle size. Particles with smaller diameters travel more quickly, while those with greater diameters move slower. Therefore, DLS offers a measure of particle hydrodynamic diameter based on the variance in the scattered laser light's intensity variation. The Stokes-Einstein equation is used by DLS to estimate size[1]:

$$d = \frac{k_B T}{3\Pi\eta D}$$

where k_B—Boltmann Constant,
T—temperature,
H—viscosity,
D—diffusion coefficient.

This method is well-suited for particle sizing because of the enormous range of submicron particle sizes it can reach and the velocity with which it collects data. The time-dependent fluctuation in particle size distribution is verified for stable and unstable nanofluids with various particle loads. After sonication, unstable nanofluids demonstrate an increase in particle size, while stable nanofluids confirm no change in particle size. Nanofluids comprising copper oxide, titanium dioxide, and single-wall carbon nanotubes in water were studied for stability using DLS. The dispersion stability of titanium and alumina nanofluids based on propylene glycol was examined using visual inspection and particle size analysis. Aqueous silica NPs' stability can be determined using DLS[6-8].

The stability of Al_2O_3 nanofluid is determined by measuring the size of nanoclusters as a function of ultrasonic mixing time for varied concentrations of nanofluid. DLS techniques were used to quantify particle size distributions for various concentrations of kerosene-based alumina nanofluids and found to be extremely close to the nominal particle size, indicating the presence of well-dispersed NPs without any agglomeration. Nanofluid stability and particle agglomeration in water-based ZnO nanofluids were measured using DLS[9-14].

Another scattering method that can be used to examine nanomaterials in suspension is Small Angle X-ray Scattering (SAXS). Over the years, SAXS has shown to be an effective method for determining whether or not solutions contain aggregates. There are significant variations from the pure particle scattering function due to aggregate scattering, especially at low momentum transfer rates. Small angle scattering studies are used to analyze the particle size distribution and aggregate structure of various ferro fluids. The aggregate structure in ferro fluids utilizing SAXS have been made on samples with chain-like particle order under a strong magnetic field. DLS and SAXS measurements were used to examine the particle size distribution of a stable silicon carbide nanofluid.

The TEM and SAXS results provide the information about the samples of interacting spheres with particle size distributions. It is possible to estimate sample polydispersity by comparing samples' SAXS data with the Porod region fitted to an electron density distribution smeared by Gaussian distributions of sphere diameter. With the aid of SAXS, the polydispersity of low-refraction-index perfluorinated polymer colloids was determined, compared to results from DLS. Gold NP nucleation and growth were studied with SAXS and UV-visible spectroscopy. Quantitative analysis of the SAXS data on an absolute scale reveals the size distribution, number density, and yield in real time. SAXS was used to describe Fe_3O_4 and CuO NPs of various sizes and surface morphologies, and the results were compared to TEM and DLS[15–18].

11.2.2 ZETA POTENTIAL MEASUREMENTS

The electrokinetic properties of a NP have a direct effect on its stability in aqueous solution. To generate strong repulsive forces, stable nanofluids with a high surface charge density on the NP surface are commonly used. Electrophoretic behavior is essential for understanding nanofluid dispersion; hence zeta potential measurements are used to examine electrophoretic activity. A zeta potential analyzer can be used to determine the surface charge of NPs in a solution. Zeta potential is the difference in potential between the dispersion medium and the fluid layer connected to the particle. According to dispersion theory, it measures the repulsion between nearby, similarly charged particles. There are two types of colloids: those with high zeta potential (positive or negative in the range of 40–60 mV) and those with low (negative) potential. An abundance of negative zeta potential in water gives gold nanofluids good stability.

Water-based nanofluids were tested for their stability using a zeta potential analyzer to determine the effects of pH and sodium dodecylbenzene sulfonate (SDBS). Fourier transform infrared spectra were used to examine the absorption mechanisms of surfactants on MWNT surfaces using zeta potential measurements. Zeta potential was used to assess the stability of aqueous silica nanofluids. Water-based Cu nanofluid dispersion behavior under varied pH values, dispersant types, and concentrations were studied using zeta potential. Zeta potential measurements were used to determine the water-based gold nanofluid's stability without surfactants. The zeta size distribution of water-based CNT was used to investigate the colloidal stability of the CNTs. Zeta potential was used to study the stability of water-based CNTs in the presence of functionalized group concentration.

Zeta potential measurements were used to determine the stability of aqueous metal oxide nanofluids. Ethylene glycol-based CuO nanofluids were tested for their stability by measuring their zeta potential. Zeta Sizer was used to measure the zeta potential of the CTAB-stabilized CuO nanofluid to assess its dispersion stability. Anatase and Rutile samples in aqueous dispersion were found to be stable in neutral and alkaline pH ranges due to electrostatic stabilization, which was explained in terms of DLVO theory[16–19].

11.2.3 SEDIMENTATION AND CENTRIFUGATION METHOD

Sedimentation is the most straightforward approach for determining the stability of a nanofluid. The stability of the nanofluid is indicated by the volume or quantity of sediment that accumulates in the system. When the concentration or size of the supernatant particles is constant throughout time, the nanofluids are assumed stable. The stability of graphite suspensions and water-based alumina nanofluids can be measured using the sedimentation balance method, which employs sedimentation techniques. The stability of suspensions is appealing colloidal particles by recording photographs of emulsions as they cream and measuring the position of the interface between the emulsion and the clear fluid supernatant.

The stability of aqueous copper nanofluid is studied using sedimentation techniques. Centrifugation is typically used to examine the stability of nanofluids because sedimentation is a time-consuming process. The stability of aqueous polyaniline colloids can also be determined using this approach. Polymer-coated magnetic NPs' centrifugation-based clustering and stability were both investigated[20–24].

11.2.4 SPECTRAL ANALYSIS

For nanofluid stability, spectral analysis is another effective method. A relationship exists between the amount of absorbance and the amount of NPs in the fluid. An absorption band with characteristic absorption characteristics in the wavelength range between 190 and 1100 nm can be used to determine the stability of nanofluids using UV-visible spectral analysis. The stability of alumina and copper nanofluids was examined using the sedimentation method and the absorbency analysis. Analysis of the spectrophotometer showed that FePt nanofluids were stable. Investigating nanofluid's absorbance provided information on sedimentation kinetics. The UV-visible absorption of MWNT nanofluids was measured using a UV-visible spectrophotometer to determine the stability of the nanofluids.

The absorbance method was used to investigate the stability of water-based Al_2O_3 and SiO_2 nanofluids. The change in optical absorbance of suspensions measured using a UV-visible spectrophotometer was used to determine the titanium dioxide aqueous suspension colloidal stability. In order to determine the stability of alumina nanofluid, UV absorbency spectra were used. The time-dependent thermal conductivity of nanofluid can be used to gauge its stability. Stable nanofluid's k was shown to be time independent, whereas unstable nanofluid's k was time-dependent.

The time-dependent changes in beam absorbance were used to study the magnetic suspensions' colloidal stability. The optical absorbance of suspensions of magnetite, polylactic acid (PLA), and a composite core/shell was measured over time to determine the aggregation and sedimentation of the suspensions. An MWCNT suspension in water was examined for stability using visual inspection and UV visible absorbance. The presence of agglomeration in nanofluids can be indirectly confirmed using a variety of techniques, including the 3ω method, TEM, SEM, STEM, and light scattering[25-28].

11.3   STABILITY MECHANISM OF NANOFLUID

The tendency of NPs to agglomerate into bigger aggregates results in sedimentation/ phase separation. A particle's stability in solution is governed by the sum of van der Waals attraction and electrical double layer repulsive forces, according to DLVO theory. Particles will aggregate and settle in a lower energy state if there is insufficient resistance to draw the attractive van der Waals forces to prevent the system from reaching this state. The suspensions will stay stable and free of aggregates if the particles exhibit a net repulsive force. Steric, electrostatic, and electro-steric forces can be used to keep scattered particles from colliding with each other. These NPs have an extra steric repulsive force due to the addition of polymers to the NPs' surface.

Surface charge is developed on NPs in the case of electrostatic stabilization[29]:

1. By preferential adsorption of ions, by dissociation of surface charged species.
2. By isomorphic substitution of ions.
3. By accumulation or depletion of electrons at the surface.
4. By physical adsorption of charged species onto the surface. Steric effect is used to create stable graphite nanofluids using poly vinyl alcohol (PVP).

The ZnO nanofluid stabilized by PMMA has a superior stability. Nanofluids' stability can be improved in a number of ways. A nanofluid's stability can be improved by adding surfactants. NPs become moister when surfactant is added because the amphiphilic molecules of surfactants have hydrophilic as well as hydrophobic portions, lowering the base fluid surface tension. Surfactants prefer to locate at the interface between the NP and the base fluid, where they introduce a degree of continuity between the NPs and the fluids.

Water-based Fe_3O_4 nanofluids can be generated by stabilizing the magnetic NPs using surfactant double layer. As a result of the co-precipitation method used to prepare the ultra-fine super paramagnetic magnetite, a mixture of surfactants including dodecylbenzenesulfonic acid, dodecanoic acid, myristic acid, tetradecanoic acid, oleic acid, and sodium alginate were applied on top of the magnetite to enhance its magnetic properties. The water-based alumina and copper nanofluids were prepared by directly mixing NPs with water solution and SDBS surfactant, and then ultrasonication for 1hour at a frequency of 40 kHz. As a stabilizer, CTAB was used in a two-step method to produce stable water-based CuO. The CuO NPs were first produced using a wet chemical reduction technique and dispersed in water using CTAB as a surfactant[30].

It was possible to create stable water-based TiO_2 nanofluids by stabilizing the NP using poly (acrylic acid). For 24 hours, the mixture was agitated at room temperature to allow adsorption to take place. It was found that the surfactant, sodium dodecyl sulphate (SDS), and polyvinyl pyrrolidone (PVP), play an essential role in dispersing the NPs into the base fluid as well as enhancing the stability of Al_2O_3/deionized water nanofluids. Surfactant was discovered to improve the stability of aqueous copper nanofluids made using a two-step process. PVP was used as a dispersant to make stable copper nanofluids based on ethylene glycol. Carbon black and silicon oil-based silver nanofluids were generated utilizing a two-step process with the aid of the stirrer, ultrasonic bath, and ultrasonic disruptor.

The high-pressure homogenizer was found to be the most successful method to produce exceptionally stable nanofluids. With the use of PVA as a surfactant, the stability of water-based CuO nanofluids is increased. In this study, kerosene-based nanofluids containing silver NPs coated with dithiophosphates were found to be more stable than NPs coated with other types of dithiophenes, and hence ideal for the preparation of oil-based nanofluids. Thermal breakdown of an iron organic precursor in an organic medium produced oleic acid-coated magnetite NPs with diameters ranging between 6 and 20 nm. When $FeCl_2$, $4H_2O$, and $Pt(acac)_2$ were reduced at $263°C$ in a N_2 environment, monodisperse FePt NPs were produced and dispersed in hexane with surfactants like oleic acid and alkylamine as reductants[31].

Ultrasonication for roughly 40 minutes using a 100 W, 40 kHz ultrasonicator disperses the surface-modified MWCNT in water and ethylene glycol solution. With the help of SDS, CTAB, and a dispersant, a water-based MWCNT has been created. A 750 W ultrasonic probe was used to sonicate the water-based MWCNTs and increase their stability. When alumina NPs were dispersed in water using ultrasonic and magnetic stirring, they formed stable nanofluids. The microwave-assisted chemical precipitation approach was used to produce stable Al_2O_3 and Ag nanofluids, which were then disseminated in distilled water using a sonicator. Water/EG and 0.05M SDS were used to disperse alumina, zinc-oxide, and titanium-dioxide NPs in the mixture. An ultrasonic bath for 1 h was followed by 10 h of magnetic stirring to ensure that the nanofluid was completely homogenous[32].

The water-based Au nanorods' stability was improved by dispersing them in CTAB and then ultrasonically processing them. Stable Fe_3O_4 nanofluid was generated with SDS as a surfactant. Chemical co-precipitation was used to make surfactant-coated magnetite NPs and ultrasonication was used to disseminate them in kerosene. The magnetic nanoparticles encapsulated by oleic acid were modified by ozonolysis and the interfacial ligand exchange technique and dispersed in ethanol to form stable suspensions. To break agglomerates, commercially available Al_2O_3 and CuO NPs are combined for half an hour before being dispersed in water and ethylene glycol for 30 minutes in an ultrasonication bath.

Using butanediyl-1,4-bis(dimethylcetyl ammonium bromide) and N,N-dimethylheptylamine (C16DMA) surfactants as stabilizers, we were able to create stable oil-based nanofluids that contained hydrophobic gold NPs. The colloidal stability of nanofluids stabilized by SDBS surfactant was studied and found to be superior to that of other surfactants. Surfactants have the potential to form foams when heated. In addition, adding too much surfactant might raise the thermal resistance, limiting the amount of heat that can be transferred. Functionalized NPs are utilized to create nanofluids in order to overcome the disadvantages of surfactants. A wet mechanochemical reaction was used to synthesize stable CNT nanofluids that were free of surfactant. Chemical polarization of MWNTs was used to improve water-based MWNT dispersion[33].

Steady dispersion of SiO_2-coated MWCNT in polar liquids has been achieved. Fe nanoparticles were treated with a surface modification, alkyl ethylenediaminetriacetic acid (RED3A), via a chelating process to produce composite Fe nanofluids (CINPs). The resulting CINPs particles are subsequently ultrasonically dispersed in silicon oil. Stable ferrofluids can be made without surfactants using a straightforward method. Fe_3O_4 nanoparticles were created using the co-precipitation process and then immersed in NaOH solution. The dried particles were then dissolved in water and sonicated for 30 minutes. CeO_2 nanofluid was created by dispersing surface-modified ceria (CeO_2) nanoparticles in methonal using an ultrasonicator.

The nanofluid's electrokinetic properties have a direct impact on colloidal stability, hence increasing the pH of the nanofluids has a strong repelling effect. Water-based alumina and copper suspensions were found to be more stable when their pH was close to their optimal value, according to studies on the impact of pH on their stability. Copper nanofluid was stable at pH 9.5, while alumina nanofluid was stable at pH 8.0. Additionally, a homogenizer and an ultrasonic bath were employed to break up the clusters created due to agglomeration and to improve the stability of nanofluids. This nanofluid has a pH of 9.5 or below, which is ideal for its stability[34–39].

REFERENCES

[1] Angayarkanni, S. A. & Philip, J., 'Review on thermal properties of nanofluids: Recent developments', *Advances in Colloid and Interface Science*, vol. 225, pp. 146–176, 2015.

[2] Shima, P. D., Philip, J. & Raj, B., 'Influence of aggregation on thermal conductivity in stableand unstable nanofluids', *Appl Phys Lett*, vol. 97, p. 153113, 2010.

[3] Fedele, L., Colla, L., Bobbo, S., Barison, S. & Agresti, F., 'Experimental stability analysis of different water-basednanofluids', *Nanoscale Res Lett*, vol. 6, pp. 1–8, 2011.

[4] Palabiyik, I., Musina, Z., Witharana, S. & Ding, Y., 'Dispersion stability and thermal conductivity of propylene glycol-based nanofluids', *J Nanopart Res*, vol. 13, pp. 5049–5055, 2011.

[5] Metin, C. O., Lake, L. W., Miranda, C. R. & Nguyen, Q. P., 'Stability of aqueous silica nanoparticledispersions', *J Nanopart Res*, vol. 13, pp. 839–850, 2011.

[6] Sadeghi, R., Etemad, S. G., Keshavarzi, E. & Haghshenasfard, M., 'Investigation of alumina nanofluid stability by UV-vis spectrum', *MicrofluidNanofluid*, vol. 18: 1023–1030, 2015.

[7] Agarwal, D. K., Vaidyanathan, A. & Kumar, S. S., 'Synthesis and characterization ofkerosenealuminananofluids', *ApplThermEng*, vol. 60, pp. 275–284, 2013.

[8] Niyaghi, F., Haapala, K. R., Harper, S. L. & Weismiller, M. C., 'Stability and biological responses of Zinc Oxide Metalworking Nanofluids (ZnOMWnF™) using dynamic light scattering and zebrafish assays', *Tribol Trans*, vol. 57, pp. 730–739, 2014.

[9] Eberbeck, D. & Blasing, J., 'Investigation of particle size distribution and aggregate structure of various ferrofluids by small-angle scattering experiments', *J ApplCrystallogr*, vol. 32, pp. 273–280, 1999.

[10] Bläsing, J., Strassburger, G. & Eberbeck, D., 'Determination of particle size distributionand correlation of particles in ferrofluids under the influence of magnetic fields', *Phys Status Solidi A*, vol. 146, pp. 595–602, 1994.

[11] Balasoiu, M., et al., 'Structural studies of ferrofluids by small-angle neutron scattering', *Magnetohydrodynamics*, vol. 40, pp. 421–430, 2004.

[12] Singh, D., et al., 'An investigation of silicon carbide-water nanofluid for heat transferapplications', *J Appl Phys*, vol. 105, p. 064306, 2009.

[13] Rieker, T., Hanprasopwattana, A., Datye, A. & Hubbard, P., 'Particle size distribution inferred from small-angle X-ray scattering and transmission electron microscopy', *Langmuir*, vol. 15, pp. 638–641, 1999.

[14] Wagner, J., Hartl, W. & Hempelmann, R., 'Characterization of monodisperse colloidal particles: Comparison between SAXS and DLS', *Langmuir*, vol. 16, pp. 4080–4085, 2000.

[15] Abecassis, B., Testard, F., Spalla, O. & Barboux, P., 'Probing in situ the nucleation and growth of gold nanoparticles by small-angle X-ray scattering', *Nano Lett*, vol. 7, pp. 1723–1727, 2007.

[16] Gopinath, S. & Philip, J., 'Preparation of metal oxide nanoparticles of different sizes andmorphologies, their characterization using small angle X-ray scattering and studyof thermal properties', *Mater Chem Phys*, vol. 145, pp. 213–221, 2014.

[17] Kim, H. J., Bang, I. C. & Onoe, J., 'Characteristic stability of bare Au-water nanofluids fabricated by pulsed laser ablation in liquids', *Opt Lasers Eng*, vol. 47, pp. 532–538, 2009.

[18] Wang, X. J., Li, X. & Yang, S., 'Influence of pH and SDBS on the stability and thermal conductivity of nanofluids', *Energy Fuels*, vol. 23, pp. 2684–2689, 2009.

[19] Zhu, D., Li, X., Wang, N., Wang, X., Gao, J. & Li, H., 'Dispersion behavior and thermal conductivity characteristics of Al2O3—H2O nanofluids', *CurrAppl Phys*, vol. 9, pp. 131–139, 2009.

[20] Chen, L. & Xie, H., 'Properties of carbon nanotube nanofluids stabilized by cationic gemini surfactant', *Thermochim Acta*, vol. 506, pp. 62–66, 2010.

[21] Li, X., Zhu, D. & Wang, X., 'Evaluation on dispersion behavior of the aqueous coppernano-suspensions', *J Colloid Interface Sci*, vol. 310, pp. 456–463, 2007.

[22] Nasiri, A., Niasar, M. S., Rashidi, A., Amrollahi, A. & Khodafarin, R., 'Effect of dispersion method on thermal conductivity and stability of nanofluid', *ExpTherm Fluid Sci*, vol. 35, pp. 717–723, 2011.

[23] Talaei, Z., Mahjoub, A. R., Rashidi, A. M., Amrollahi, A. & Meibodi, M. E., 'The effect of functionalized group concentration on the stability and thermal conductivity of carbonnanotube fluid as heat transfer media', *IntCommun Heat Mass Transfer*, vol. 38, pp. 513–517, 2011.

[24] Zhao W. L., Zhu, B. J., Li, J. K., Guan, Y. X. & Li, D. D., 'Suspension stability and thermal conductivity of oxide based nanofluids with low volume concentration', *Adv Mater Res*, vol. 160–162, pp. 802–808, 2011.

[25] Souza, N. S., et al., 'Stability issues and structure-sensitive magnetic properties of nanofluid ferromagnetic graphite', *J Nanofluids*, vol. 1, pp. 143–147, 2012.

[26] Sahooli, M. & Sabbaghi, S., 'CuONanofluids: The synthesis and investigation of stabilityand thermal conductivity', *J Nanofluids*, vol. 1, pp. 155–160, 2012.

[27] Anandan, D. & Rajan, K. S., 'Synthesis and stability of cupric oxide-based nanofluid: Anovel coolant for efficient cooling Asian', *J Sci Res*, vol. 1454, pp. 1–10, 2012.

[28] Eremenko, B. V., Bezuglaya, T. N., Savitskaya, A. N., Malysheva, M. L., Kozlov, I. S. & Bogodist, L. G., 'Stability of aqueous dispersions of the hydrated titanium dioxide prepared by titanium tetrachloride hydrolysis', *Colloid J*, vol. 63, pp. 173–178, 2001.

[29] Wei, X. & Wang, L., 'Synthesis and thermal conductivity of microfluidic coppernanofluids', *Particuology*, vol. 8, pp. 262–271, 2010.

[30] Angayarkanni, S. A. & Philip, J., 'Effect of nanoparticles aggregation on thermal and electrical conductivities of nanofluids', *J Nanofluids*, vol. 3, pp. 17–25, 2014.

[31] Nasiri, A., Shariaty-Niasar, M., Rashidi, A. M. & Khodafarin, R., 'Effect of CNT structures onthermal conductivity and stability of nanofluid', *Int J Heat Mass Transfer*, vol. 55, pp. 1529–1535, 2012.

[32] Kim, C., et al., 'Gravitational stability of suspensions of attractive colloidal particles', *Phys Rev Lett*, vol. 99, pp. 028303, 2007.

[33] Li, D. & Kaner, R. B., 'Processable stabilizer-free polyaniline nanofiber aqueous colloids', *ChemCommun*, vol. 14, pp. 3286–3288, 2005.

[34] Mali, S., Pise, A. & Acharya, A., 'Review on flow boiling heat transfer enhancement withnanofluids', *IOSR-JMCE*, vol. 11, pp. 43–48, 2014.

[35] Ditsch, A., Laibinis, P. E., Wang, D. I. C. & Hatton, T. A., 'Controlled clustering and enhanced stability of polymer-coated magnetic nanoparticles', *Langmuir*, vol. 21, p. 6006, 2005.

[36] Huang, J., Wang, X., Long, Q., Wen, X., Zhou, Y. & Li, L., 'Influence of pH on the stability characteristics of nanofluids', In: *Symposium on Photonics and Optoelectronics (SOPO '09)*, IEEE xplore, 2009.

[37] Farahmandjou, M., Sebt, S. A., Parhizgar, S. S., Aberomand, P. & Akhavan, M., 'Stability investigation of colloidal FePt nanoparticle systems by spectrophotometer analysis', *Chin Phys Lett*, vol. 26, p. 027501, 2009.

[38] Hwang, Y., et al., 'Stability and thermal conductivity characteristics of nanofluids', *Thermochim Acta*, vol. 455, p. 70, 2007.

[39] Gallego, M. J. P., Casanova, C., Paramo, R., Barbes, B., Legido, J. L. & Pineiro, M. M., 'A study on stability and thermophysical properties (density and viscosity) of Al_2O_3 in waternanofluid', *J Appl Phys*, vol. 106, p. 064301, 2009.

12 Effect of Particle Volume Concentration

CONTENTS

12.1   INTRODUCTION

Figure 12.1 shows the effect of particle volume concentration on nanofluid thermal conductivity enhancement for Al_2O_3 in water from the works of seven experimental groups. Thermal conductivity is improved with an increased volume concentration of particles, as seen in Figure 12.2, although particle size and nanofluid temperature vary among groups. Viscosity rises of up to 1.3% (30%) are usual for oxide-particle volume concentrations below 5%.

Figure 12.2 shows the influence of particle volume concentration on thermal conductivity enhancement by comparing the data of two study groups using the identical nominal size particles. According to the available data, the two groups in Figure 12.2 employed the same parameters, and the results are practically equal. (The amount of the improvement in Figure 12.2 is very low due to the relatively small particle diameter.) The same patterns were found with various particle and fluid combinations and other particle sizes.

Figure 12.3 depicts the improved thermal conductivity of CuO in water, as measured by a variety of researchers. Particle sizes and fluid temperatures are shown in a wide range in Figure 12.3, just as they were in Figure 12.1. Figure 12.4 shows the volume concentration of CuO for a single particle size and a single fluid temperature. Figures 12.1–12.4 show the same overall trend, and the magnitudes in Figures 12.2 and 12.4 are validated by two separate groups of experimenters.

Figure 12.5 shows that the effect of particle volume concentration on ethylene glycol as the base fluid follows the same trend as the water data in Figures 12.1–12.4. The data from the two research groups in Figure 12.5 have a high degree of

DOI: 10.1201/9781003163633-12

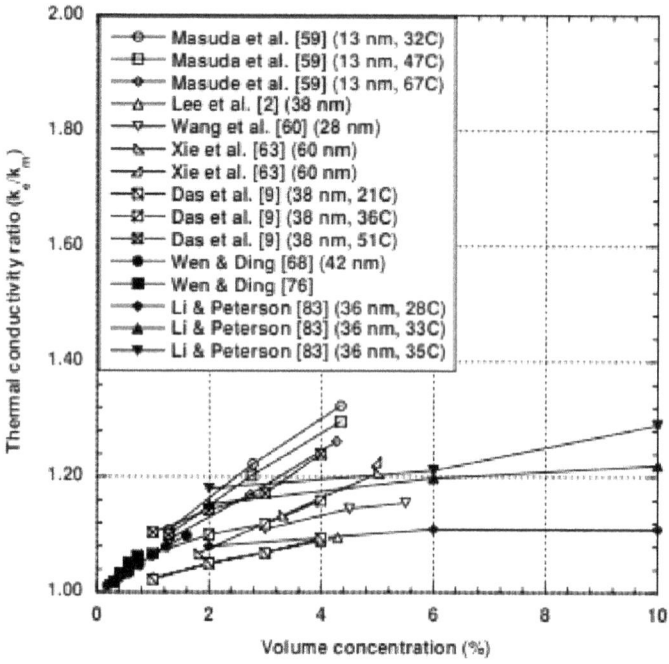

FIGURE 12.1   Thermal conductivity enhancement of Al_2O_3 in water[1-3].

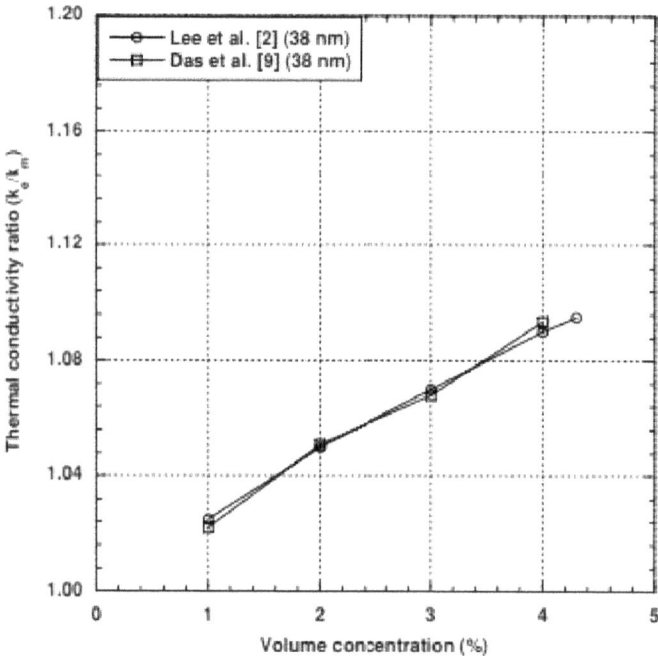

FIGURE 12.2   Effect of particle volume concentration for Al_2O_3 in water[1-3].

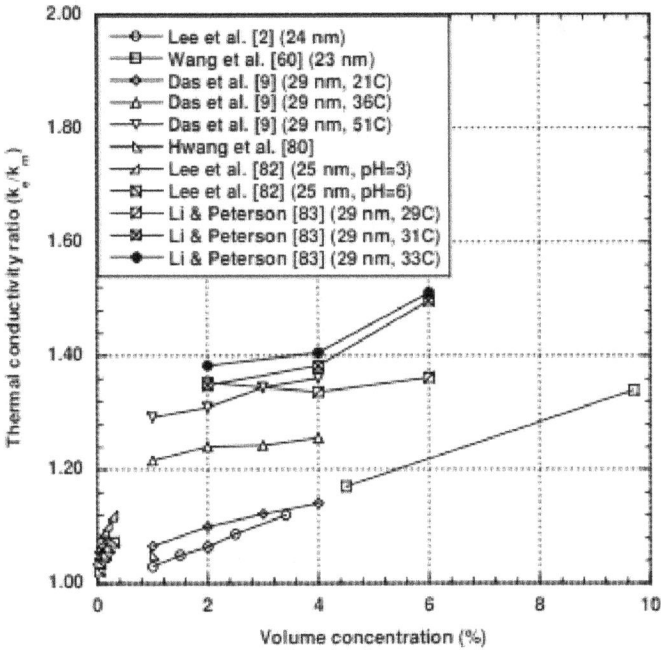

FIGURE 12.3 Thermal conductivity enhancement of CuO in water[1–3].

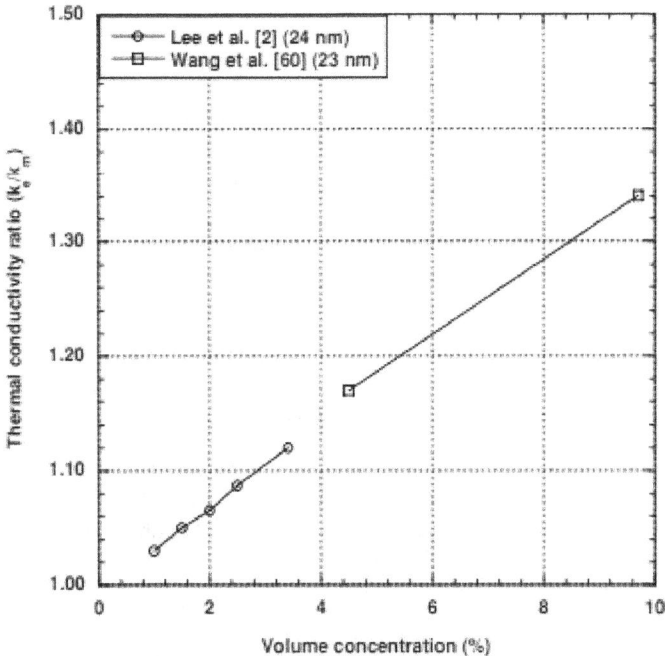

FIGURE 12.4 Effect of particle volume concentration for CuO in Water[1–3].

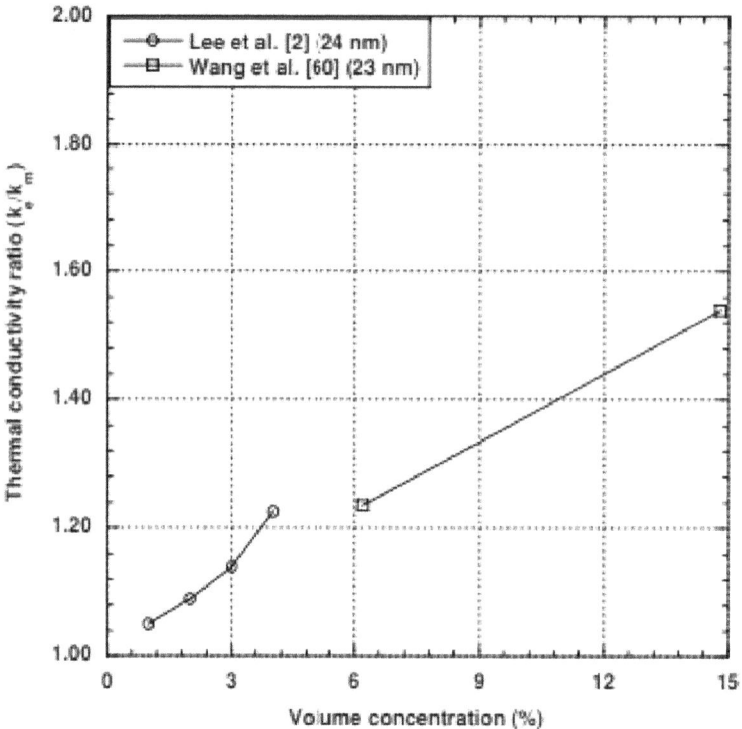

FIGURE 12.5   Effect of particle volume concentration for CuO in ethylene glycol[1–3].

agreement in magnitude. Figures 12.1–12.5 show a general rise in thermal conductivity with increasing particle volume concentration. Comparisons include several research and experimental groups agreeing on the extent of enhancement with supposedly equal test conditions for Al_2O_3, CuO, and ethylene glycol for Al_2O_3, CuO, and Al_2O_3. Nanofluid viscosity is projected to increase as the particle volume concentration increases. There were, however, no studies that refuted the trend in the feasible volume concentration range for engineering applications depicted in the Figures 12.1–12.3.

12.2   EFFECT OF PARTICLE MATERIAL

Figure 12.6 depicts the impact of particle material on thermal conductivity enhancement for two oxide particles and silicon carbide in water. Figure 12.6 shows only the effect of the material attribute, with the rest of the factors nearly constant. Low-temperature particles' thermal conductivity appears to be unaffected by their particle composition.

An example of thermal conductivity data that compares two metal particles with an oxide particle is shown in Figure 12.7. When compared to an oxide particle, metal

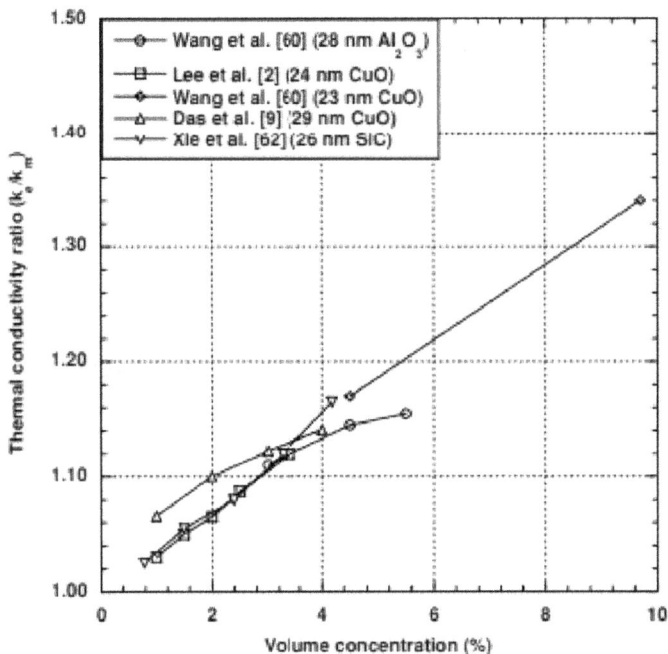

FIGURE 12.6   Effect of particle material for particles in water[3].

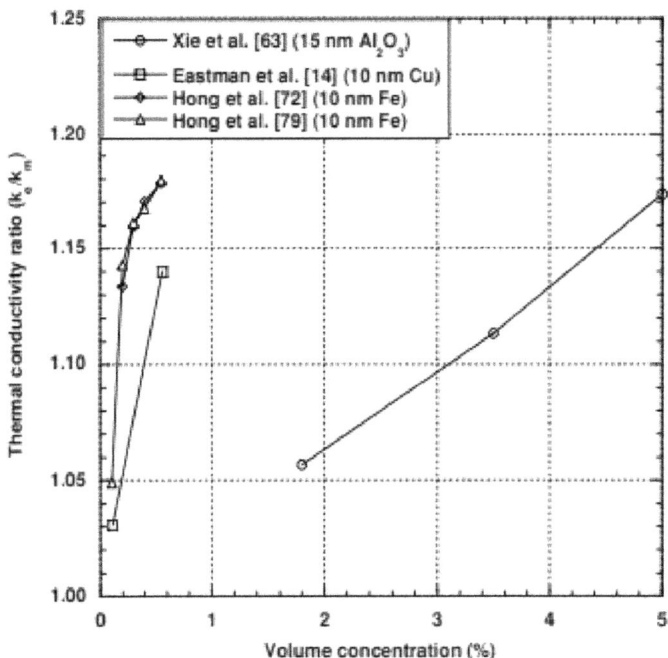

FIGURE 12.7   Effect of particle material for particles in ethylene glycol[3].

particles create a similar enhancement, but at a drastically reduced volume concentration (as indicated). On account of their lower thermal conductivity than oxide particles, metal particles were expected to perform better than oxide particles in this experiment. Creating metal particle nanofluids is difficult, because doing so risks oxidation of the NPs. Figure 12.7 shows the one-step procedure used to create the Cu particles[3].

As shown in Figure 12.8, the metal-particle nanofluid's thermal conductivity was considerably improved when the particle volume concentration was increased to 2.5% compared to the maximum particle concentration of around 0.7% in Figure 12.7. Figure 12.8 indicates that at a metal particle volume concentration of 2.5%, the nanofluid has a thermal conductivity that is 115% more than that of ethylene glycol. However, as previously mentioned, a major issue for metal-particle nanofluids is eliminating the oxidation process both during synthesis and thereafter during use. Particle coating has received some attention as a possible solution to this problem[3].

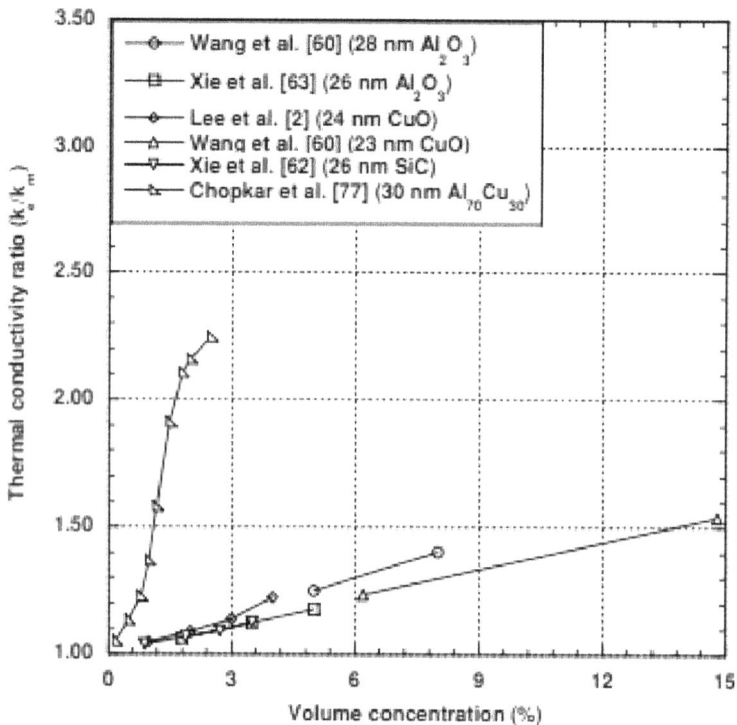

FIGURE 12.8   Effect of particle material for particles in ethylene glycol[3].

12.3   EFFECT OF PARTICLE SIZE

The nominal diameter of only spherical particles was used as the size parameter. Figure 12.9 shows the results for a single particle/water combination ranging from 28 to 60 nm in diameter. The results reveal an increase in thermal conductivity enhancement for 60 nm-sized particles. The smallest particles would have the least enhancement based on this section of the results. However, the results for the 28 nm particles fall somewhere in between the two bigger sizes. Figure 12.9 shows a rather narrow range of particle sizes for which data are available with all other factors the same, and the results are unconvincing. Figure 12.10 shows data for ethylene glycol with a limited particle size range, as well. In this case, the thermal conductivity of the intermediate-sized particle is the most improved[3].

Heat conductivity increases as the particle diameter increases, as shown in Figures 12.9 and 12.10. Certain assumptions suggest that a homogeneous distribution of small particles would be most beneficial. For thermal conductivity studies, agglomeration is a key factor, but the particle size impacts may not be as well

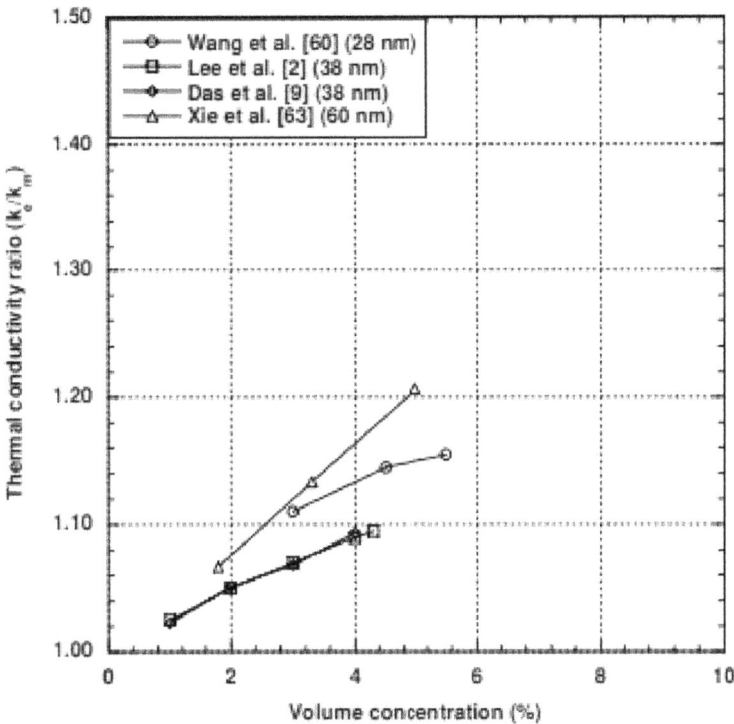

FIGURE 12.9   Effect of particle size for Al_2O_3 in water[3].

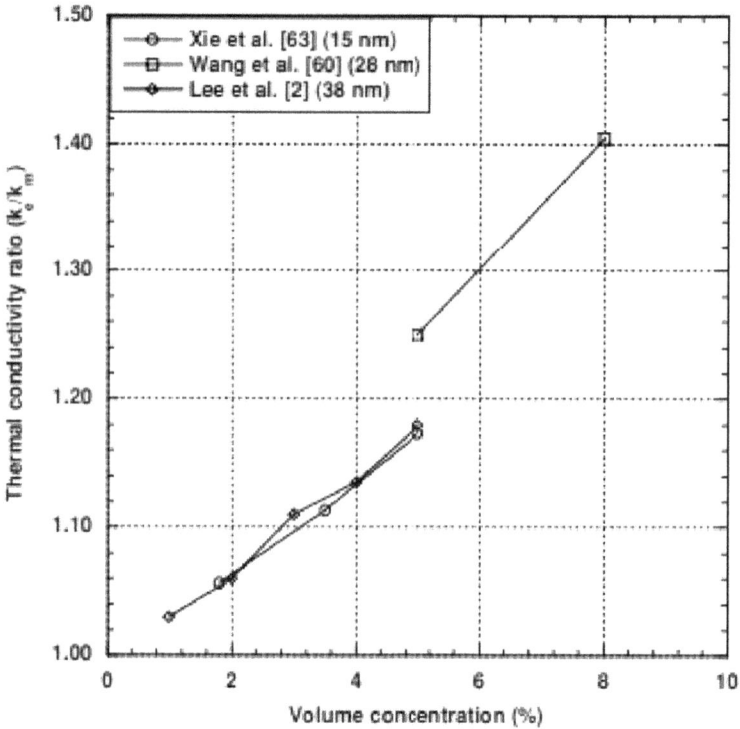

FIGURE 12.10   Effect of particle size for Al$_2$O$_3$ in ethylene glycol[3].

understood. In this situation, as in all other nanofluid data comparisons in this study, there is uncertainty due to the stated particle size and shape[3].

Figure 12.11 shows a third comparison of the influence of particle size on thermal conductivity enhancement for CuO in water, as demonstrated. While the other two groups indicate a rise in enhancement with increasing particle size, here the results are distinct. Figures 12.9–12.11 show that the thermal conductivity of suspended spherical nanometer particles increases with particle diameter[3].

12.4   EFFECT OF PARTICLE SHAPE

The geometric shape of the nanofluid particles was compared to the thermal conductivity of the nanofluids. Figure 12.12 shows a comparison of spherical and cylindrical particles. Elongated particles, which are assumed to be responsible for increasing thermal conductivity by meshing with the fluid, have been found in the cylinder samples. Data from a single group is provided in Figure 12.12, while results from other groups are shown in Figures 12.13 and 12.14. Figures 12.12–12.14, on the other hand, show that elongated particles outperform spherical ones in terms of improving thermal conductivity.

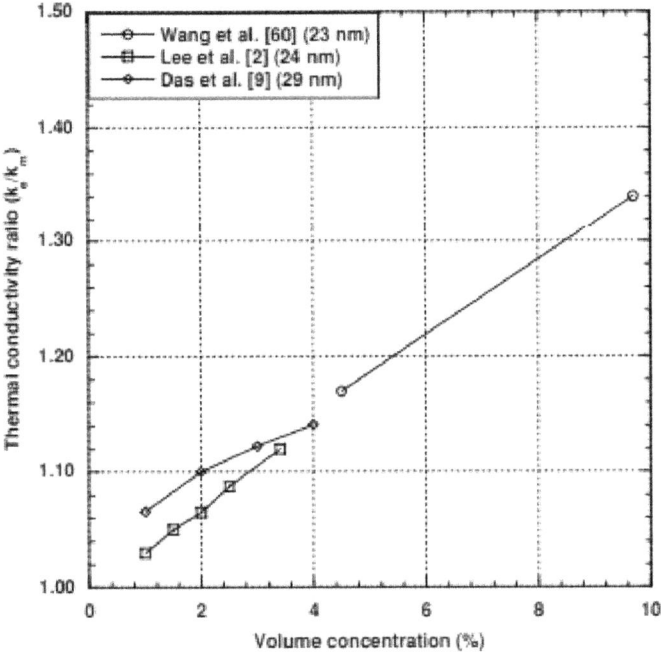

FIGURE 12.11   Effect of particle size for CuO in water[3].

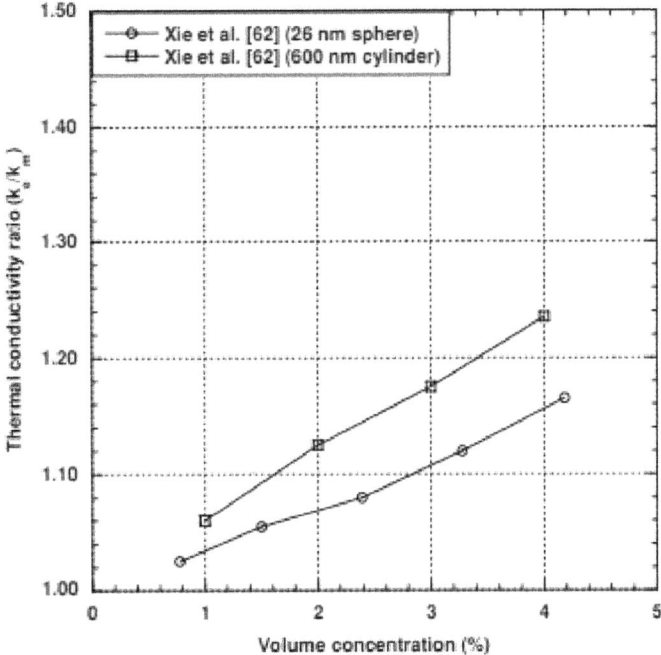

FIGURE 12.12   Effect of particle shape in SiC in water[2, 4].

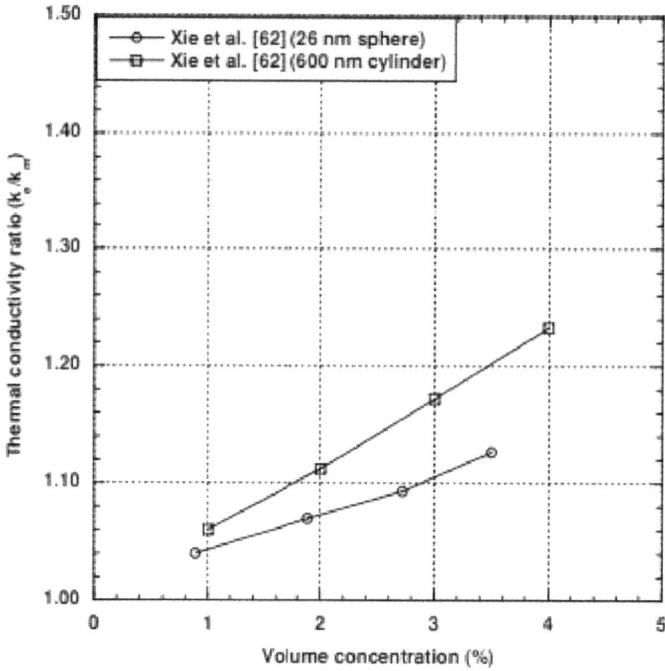

FIGURE 12.13 Effect of particle shape for SiC in ethylene glycol[2, 4].

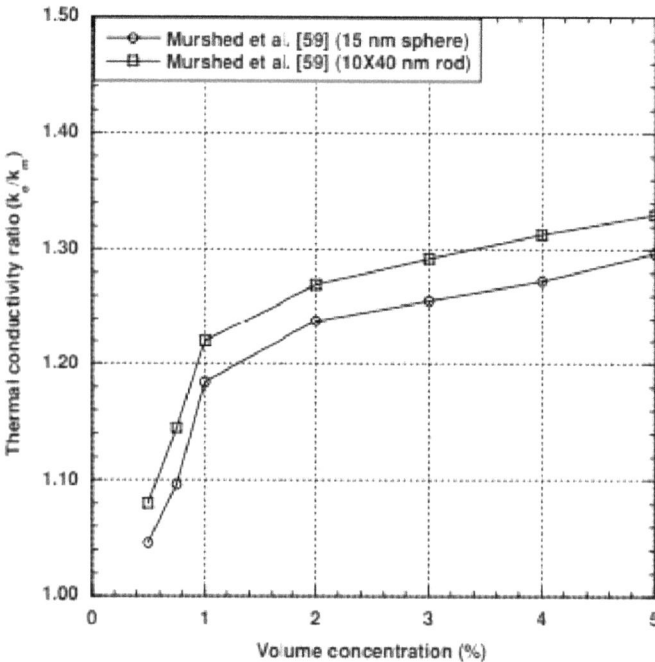

FIGURE 12.14 Effect of particle shape in SiC in water[2, 4].

12.5   EFFECT OF BASE FLUID MATERIAL

Figure 12.15 shows the effect of the base fluid on the improvement of nanofluids' thermal conductivity for three different base fluids (e.g. water, ethylene glycol, and pump oil).

Thermal conductivity improvements have a greater impact on the performance of less efficient heat transfer fluids. Figure 12.15 shows that water, the heat transfer fluid with the highest thermal conductivity, has the least enhancement. However, this overall trend was not validated by all of the experimental data studied, although it was frequently the case. Water is a better heat transfer fluid than ethylene glycol, although ethylene glycol/water mixtures are in the middle of the pack when it comes to heat transfer[2, 4].

12.6   EFFECT OF TEMPERATURE

The thermal conductivity of nanofluids is generally higher than that of the basic fluid. As a result, the thermal conductivity improvement of nanofluids is likewise

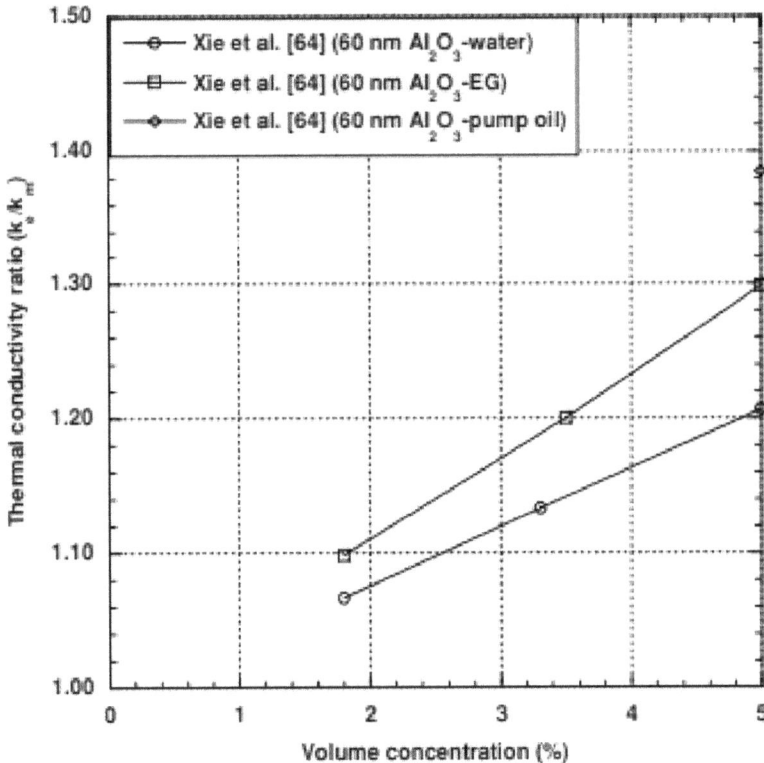

FIGURE 12.15   Effect of base fluid material for Al_2O_3 in fluids[2, 4].

temperature-dependent. Due to the wide range of temperatures utilized by the experimenters, this condition is inevitable. There are no direct comparisons amongst experimenters, although all but one experimental group demonstrates an increase in thermal conductivity with an increase in temperature. Al_2O_3 and CuO in water nanofluid results were reported across a narrow temperature range. The temperature dependence of nanofluid thermal conductivity has been associated to the movement of NPs.

Figures 12.16–12.18 show the results of Al_2O_3 in water from three separate experiments. Fluid temperatures and particle sizes vary among the figures but the particle size is constant in each and the temperature fluctuates.

Thermal conductivity of CuO in water was increased by two study groups as indicated in Figures 12.19 and 12.20. The results clearly show that as the temperature rises, so does the improvement in Figures 12.21 and 12.22.

Both SiO_2 and TiO_2 in water were shown to have a temperature trend that was not monotone with respect to temperature. All things considered, there appears to be strong support for the general temperature progress.

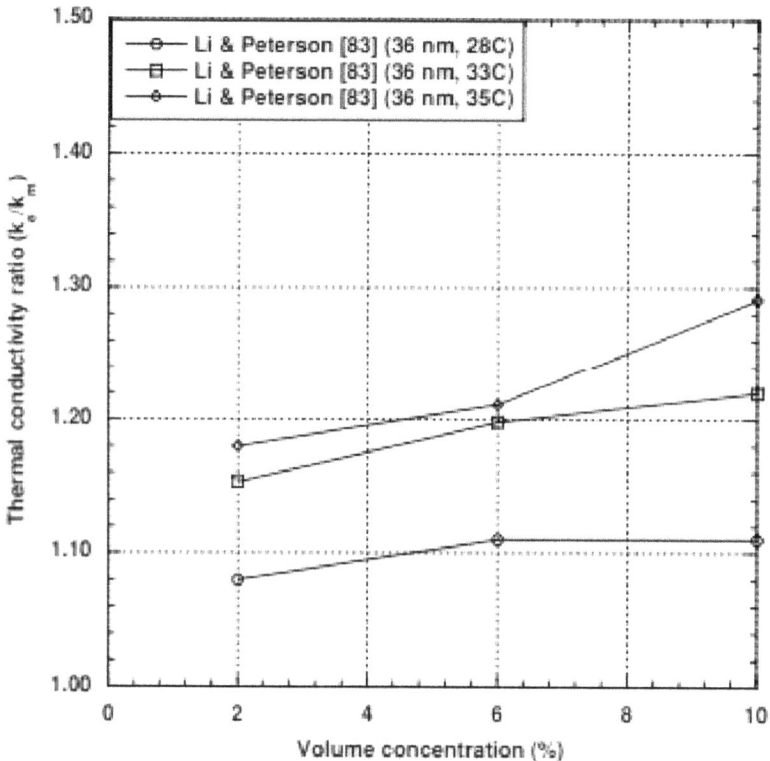

FIGURE 12.16   Effect of temperature for Al_2O_3 in water[2, 4].

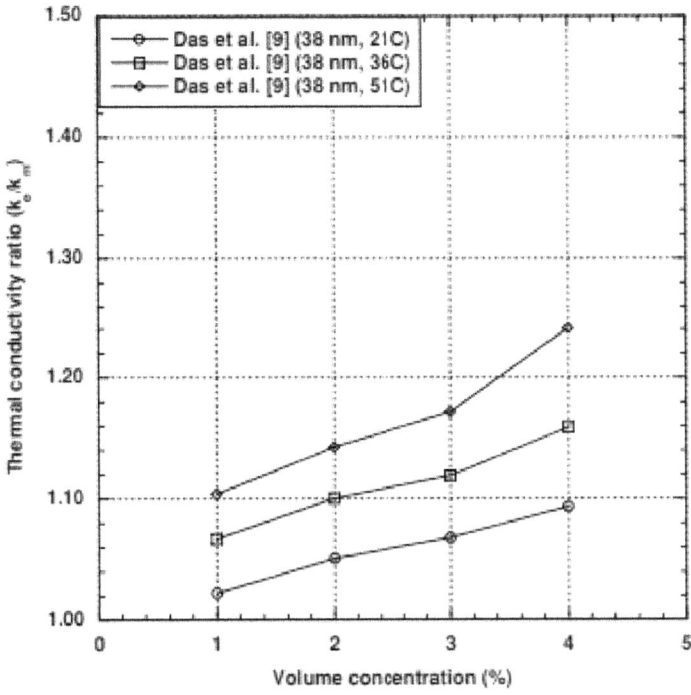

FIGURE 12.17 Effect of temperature for Al_2O_3 in water[2, 4].

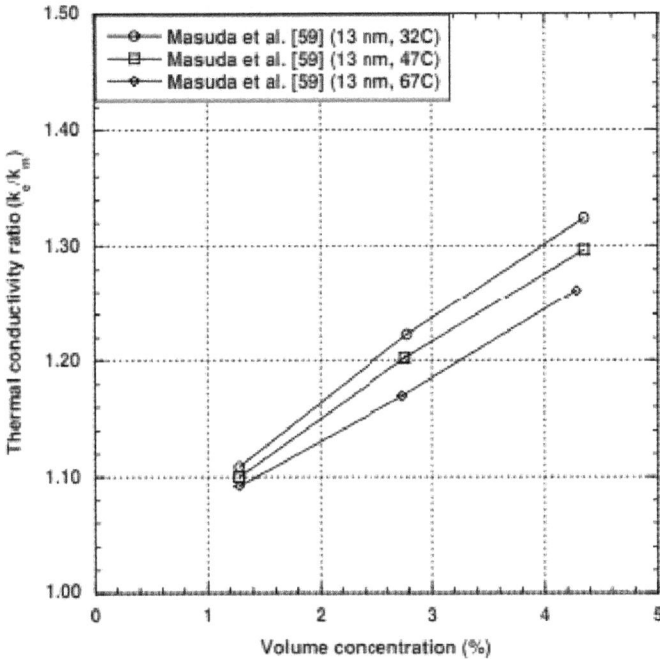

FIGURE 12.18 Effect of temperature for Al_2O_3 in water[2, 4].

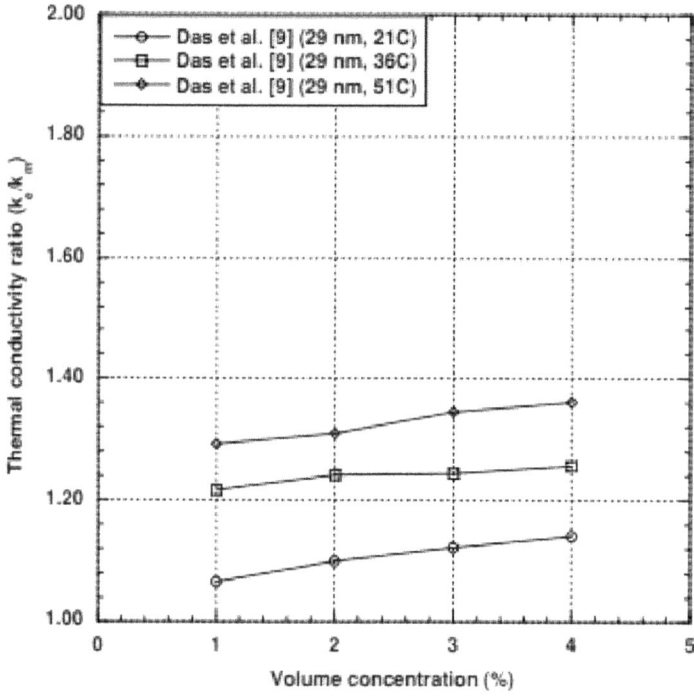

FIGURE 12.19 Effect of temperature for CuO in water[2, 4].

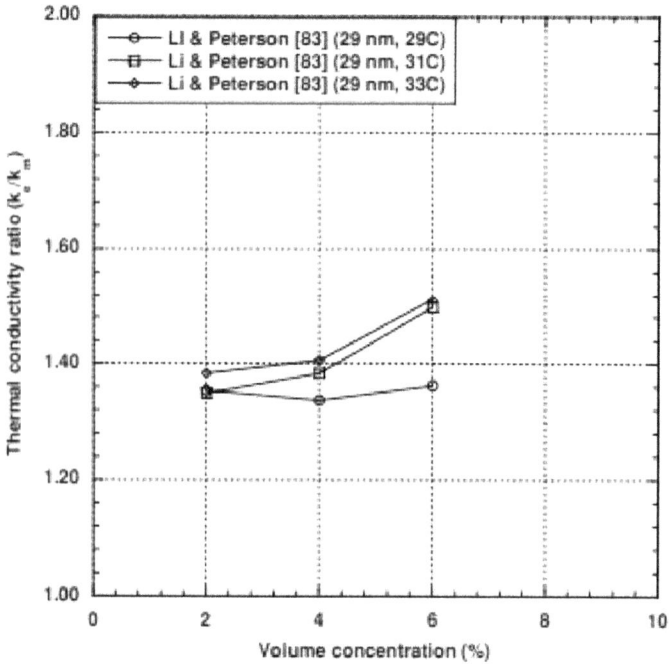

FIGURE 12.20 Effect of temperature for CuO in water[2, 4].

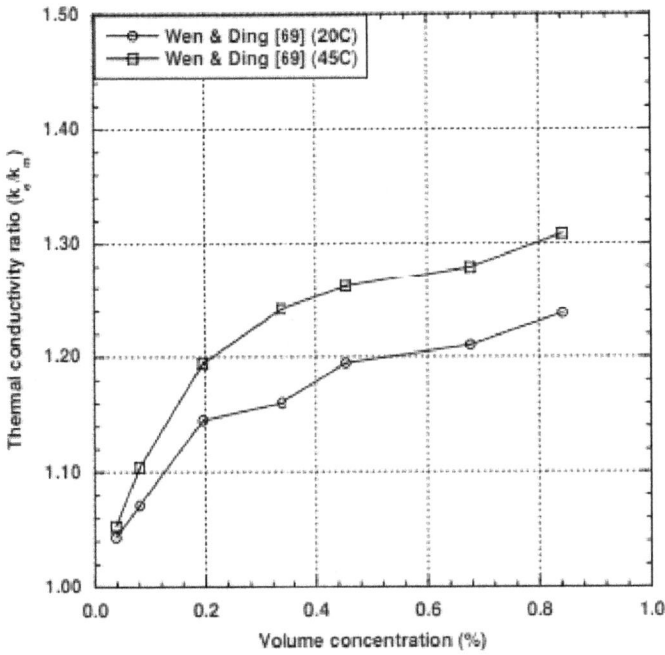

FIGURE 12.21 Effect of temperature for MWCNT in water[2, 4].

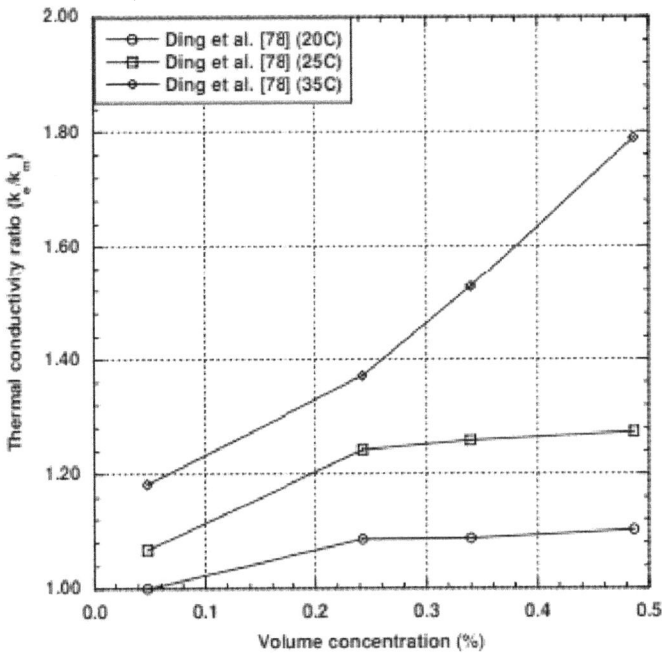

FIGURE 12.22 Effect of temperature for MWCNT in water[2, 4].

12.7   EFFECT OF ADDITIVES

Fluid additives have been employed to keep nanoparticles in suspension and prevent them from clumping together during experiments. Most experiments including additives result in an increase in the thermal conductivity ratio. This data is shown in Figures 12.23 and 12.24 for different nanofluid and additive combinations. When the additive was used in both cases, it improved the thermal conductivity enhancement.

12.8   EFFECT OF ACIDITY (pH)

Nanofluids' thermal conductivity can be improved by increasing the acidity of the fluid. Separate findings are shown for each of the two groups of participants. A similar pattern can be seen in the water particle distribution in Figures 12.25 and 12.26.

    Particle size distribution, rheology, viscosity, and stability of heat transfer nanofluids were all studied for their effects on zeta potential, pH, and a variety of other properties, including rheology, viscosity, and thermal conductivity. These findings suggest that pH affects the stability of nanofluids. Surfactants reduce particle agglomeration in heat transfer nanofluids including Fe_2O_3 and CuO NPs, according to the same group's findings in studies of alignment, pH, surfactant, and solvent impacts.

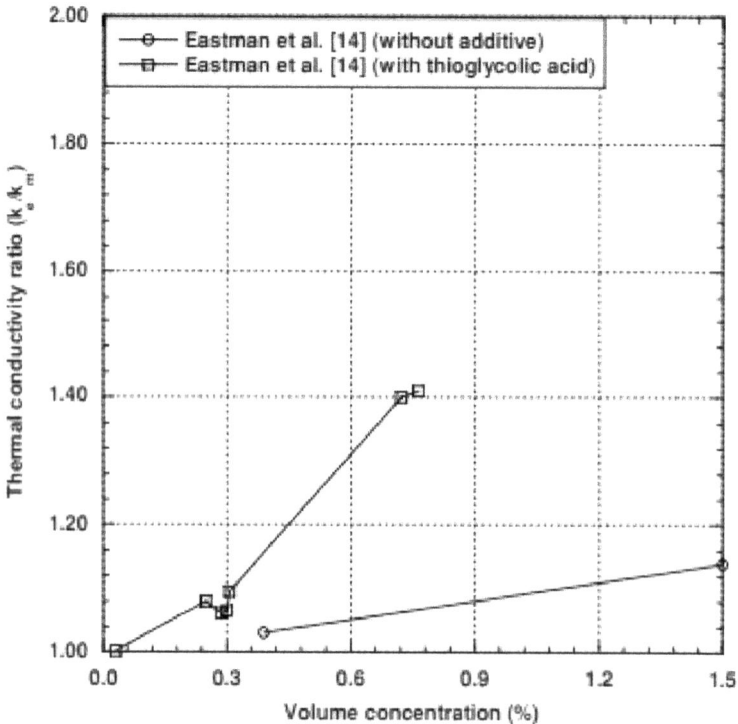

FIGURE 12.23   Effect of additive for Cu in ethylene glycol[5–7].

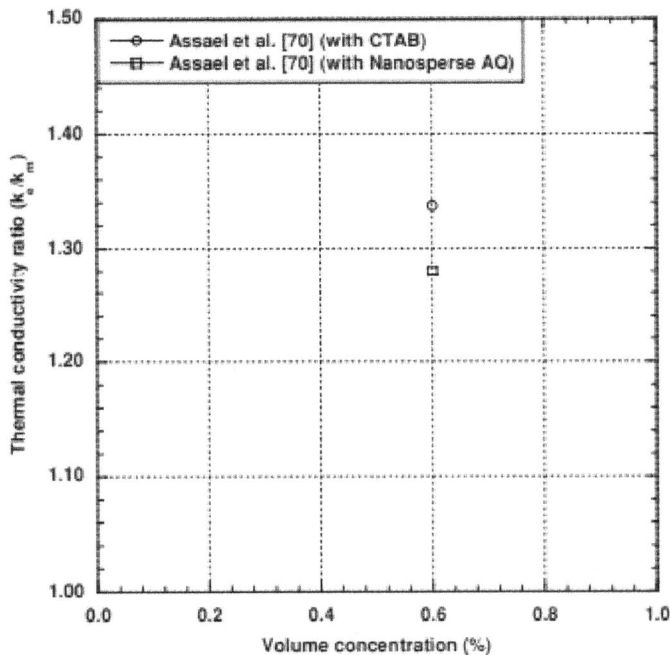

FIGURE 12.24 Effect of additive for MWCNT in water[5–7].

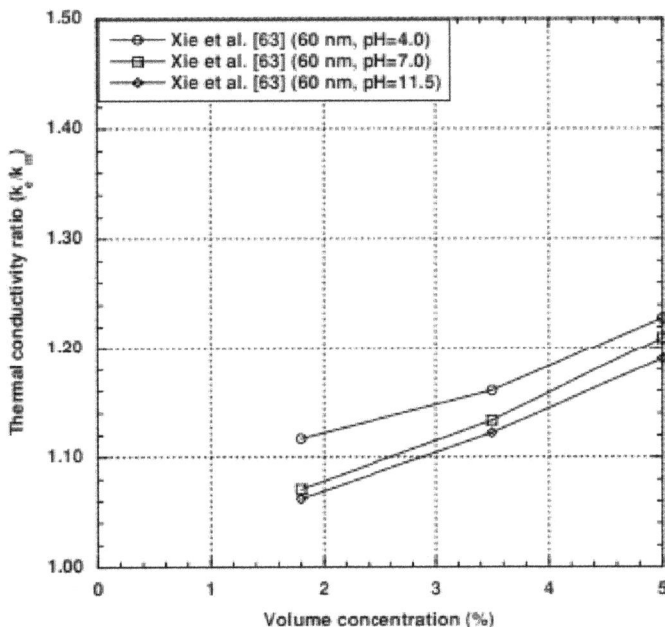

FIGURE 12.25 Effect of acidity for Al_2O_3 in water[5–7].

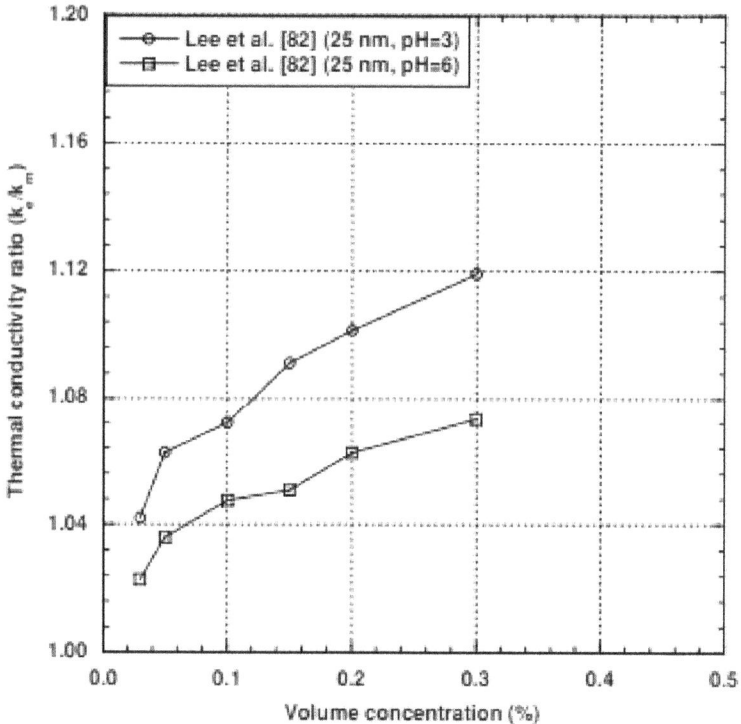

FIGURE 12.26   Effect of acidity for CuO in water[5-7].

The aggregation kinetics were manipulated by altering the pH of dispersed SiO_2 sols in water, where the clustering did not demonstrate any discernible increase in thermal conductivity even at high particle loading (23.3 vol%). The thermal conductivity ratio for different weight fractions of NPs in water-based Cu and Al_2O_3 nanofluids increased with pH at lower pH but decreased at higher pH when using these nanofluids[8-11]. It was found that nanofluids with the maximum thermal conductivity had an appropriate pH value. SnO_2 nanofluids with a pH distant from the isoelectric point (IEP) were more stable and thermally conductive when dispersed in water. Water-based Al_2O_3 nanofluid k ratio measurements with SDBS dispersant demonstrated an increase in k ratio with an increase in pH of 8.0–9.0. Water-based Al_2O_3 (1550 nm) nanofluid stability and k improvements were largely dependent on pH values, with an ideal pH value for best dispersion behavior and high k. For water-based Cu nanofluids, k was boosted by 10.7%, with an ideal pH value. For water-based nanofluids (40–50nm) a lower pH resulted in less aggregate formation, improved dispersion stability as well as a higher effective k value. More stable colloidal particles in water-based microfluids with CuO (25 nm) nanoparticles lead to a change in their k. The pH value has been shown to affect k in water-based Al_2O_3 nanofluids[12-14].

12.9 EFFECT OF SONICATION

Different sonication powers were used to study the impact of sonication on nanofluids' effective thermal conductivity (k). The smallest nanoparticle cluster size and the largest effective k enhancement were achieved with 450 W sonication powers in both alumina and copper oxide nanofluids. It has been discovered that ultrasonication has a two-fold effect on MWCNT nanofluid water. Ultrasonication promotes in dispersion formation when used below the optimal processing period; however, when used beyond the optimal time, ultrasonication increases nanotube breaking and decreases the aspect ratio of CNTs[6].

Water-based SWCNT and MWCNT nanofluid studies revealed that k of nanofluid is time-dependent immediately after ultrasonication and independent after a longer time. With increasing ultrasonication time, nanofluid k was found to increase in EG-based CNT nanofluids. Sonication has been shown to cause the formation of agglomeration of particles in water-based CuOnanofluid systems, which has been confirmed microscopically. When dispersing energy and dispersing time are increased in a poly α-olefin (PAO) based MWCNT nanofluid, the aspect ratio lowers and k decreases with increasing dispersing energy, respectively. With no surfactant, water-based Cu nanofluids exhibited time-dependent properties[7].

The length of the nanotubes in EG-based CNT nanofluids was reduced by ultrasonic homogenization, which was validated by SEM. After sonication, EG-based Fe (11 nm) nanofluids had a higher k. It took the nanofluid for 30 minutes to show 18% improvement after 30 minutes of sonication, and it was saturated after that time. Ultrasonication was shown to break the nanoclusters into smaller clusters by measuring the cluster size. k of nanofluids was discovered to be closely related to NP clustering based on time-dependent fluctuations in cluster size and k. There was an immediate 14.2% increase in k enhancement after the sonication and the increase drops to 8.5% after an hour. SEM confirmed that increasing homogenization time in water-based CNT nanofluids decreased the suspension's k due to nanotube breaking and length reduction.

12.10 EFFECT OF AGGREGATION

Stabilizing single-walled carbon nanotubes in water with magnetic NPs was possible using high-speed microscopy. When a magnetic field is applied, it has been found that the thermal conductivity of magnetic-metal-coated carbon nanotubes can considerably increase. An increase in the ratio of k/kf in a magnetic nanofluid in the presence of an external magnetic field was seen in studies. An increase in k is observed when magnetic field is applied to Fe NPs (26nm) suspended in water. In Ni and Fe_2O_3 integrated SWCNTs distributed in water, magnetic field-induced k enhancement has also been found. Suspension of MMPCM (Magnetic Microencapsulated Phase Change Material) shows an increase in k/kf when exposed to an external magnetic field. In kerosene-based Fe_3O_4 nanofluids, a recent study found that a transverse magnetic field improved k/kf [6].

REFERENCES

[1] Lee, S., Choi, S. U. S., Li, S. & Eastman, J. A., 'Measuring thermal conductivity of fluids containing oxide nanoparticles', *Transactions of the ASME, Journal of Heat Transfer*, vol. 121, pp. 280–289, 1999.

[2] Das, S. K., Putra, N., Thiesen, P. & Roetzel, W., 'Temperature dependence of thermal conductivity enhancement for nanofluids', *Transactions of the ASME, Journal of Heat Transfer*, vol. 125, pp. 567–574, 2003.

[3] Wang, X., Xu, X. & Choi, S. U. S., 'Thermal conductivity of nanoparticle-fluid mixture', *Journal of Thermophysics and Heat Transfer*, vol. 13, pp. 474–480, 1999.

[4] Masuda, H., Ebata, A., Teramae, K. & Hishinuma, N., 'Alteration of thermal conductivity and viscosity of liquid by dispersing ultra-fine particles (dispersion of γ—Al_2O_3, SiO_2, and TiO_2 ultra-fine particles)', *NetsuBussei*, vol. 7, pp. 227–233, 1993.

[5] Yu, W., France, D. M., Routbort, J. L. & Choi, S. U., 'Review and comparison of nanofluid thermal conductivity and heat transfer enhancements', *Heat Transfer Engineering*, vol. 29, no. 5, pp. 432–460, 2008.

[6] Philip, J. & Shima, P. D., 'Thermal properties of nanofluids', *Advances in Colloid and Interface Science*, vol. 183, pp. 30–45, 2012.

[7] Wang, X. J., Zhu, D. S. & Yang, S., 'Investigation of pH and SDBS on enhancement of thermal conductivity in nanofluids', *Chem Phys Lett*, vol. 470, pp. 107–111, 2009.

[8] Zhu, D., Li, X., Wang, N., Wang, X., Gao, J. & Li, H., 'Dispersion behavior and thermal conductivity characteristics of Al2O3—H2O nanofluids', *CurrAppl Phys*, vol. 9, pp. 131–139, 2009.

[9] Wamkam, C. T., Opoku, M. K., Hong, H. & Smith, P., 'Effects of pH on heat transfer nanofluids containing ZrO2 and TiO2 nanoparticles', *J Appl Phys*, vol. 109, p. 024305, 2011.

[10] Younes, H., Christensen, G., Luan, X., Hong, H. & Smith, P., 'Effects of alignment, pH, surfactant, and solvent on heat transfer nanofluids containing Fe_2O_3 and CuO nanoparticles', *J Appl Phys*, vol. 111, p. 064308, 2012.

[11] Li, X. F., Zhu, D. S., Wang, X. J., Wang, N., Gao, J. W. & Li, H., 'Thermal conductivity enhancement dependent pH and chemical surfactant for Cu-H2O nanofluids', *Thermochim Acta*, vol. 469, pp. 98–103, 2008.

[12] Lee, D., Kim, J. W. & Kim, B. G., 'A new parameter to control heat transport in nanofluids: Surface charge state of the particle in suspension', *J Phys Chem B*, vol. 110, pp. 4323–4328, 2006.

[13] Liu, M. S., Lin, M. C. C., Tsai, C. Y. & Wang, C. C., 'Enhancement of thermal conductivity with Cu for nanofluids using chemical reduction method', *Int J Heat Mass Transf*, vol. 49, pp. 3028–3033, 2006.

[14] Hong, H., Wright, B., Wensel, J., Jin, S., Ye, X. R. & Roy, W., 'Enhanced thermal conductivity by the magnetic field in heat transfer nanofluids containing carbon nanotube', *Synth Met*, vol. 157, p. 437, 2007.

13 Heat Transfer Mechanism in Nanofluid

CONTENTS

13.1 INTRODUCTION

Solid-liquid combinations are typically represented using systems that consider fluids to be continuous substances. Nonetheless, when working with low-volume nanofluids, this assumption might lead to large discrepancies in experimental results. In order to accurately depict the change in the properties of a nanofluid, it is important to look for different mechanisms to characterize the molecular interaction of NPs with the base fluid[1–5].

Nanofluids' thermal conductivity can be explained by four different methods, for example:

1. The addition of suspended NPs results in an increase in heat transfer surface area.
2. The degree of thermal conductivity enhancement relies on the surface area of the NP and the rate of collisions between them.
3. By using NPs with a higher thermal conductivity, the thermal conductivity of the material is increased.
4. An increase in the dispersion of NPs results in microturbulences along the base fluid.

As a result of these interactions, there are two models that can represent the thermal transport in nanofluids.

1. For example, in the first type of thermal transport, the thermal transport can be thought of as having a conventional part and a dynamic part, which is induced by the Brownian motion of NPs along the base fluid.
2. The Brownian motion is responsible for the movement and mixing of the nanofluid components at the microscopic level, allowing for an increase in heat transfer between the components.

The thermal conductivity can be improved in two ways by this movement. The first is a direct result of the heat being carried by the NPs as they move across space.

Secondly, micro-convection surrounding the NPs generates an indirect contribution to heat transfer[6–8].

Nanofluid thermal conductivity can be modeled using the properties that directly affect its value:

1. NP size.
2. NP volume fraction.
3. Base fluid thermal conductivity.
4. NP thermal conductivity.
5. Nanofluid temperature.

Brownian motion's influence on the nanofluid is still directly connected to other nanofluid properties, such as diffusion velocity, agglomeration of particles, and particle concentrations. Figure 13.1 depicts the trajectory of a particle changing its color in water as it moves through the medium. Every 30 s, the spots on the graph correspond to the particle's location.

Figure 13.2 presents experimental results relating to the experimental temperature achieved by increasing the thermal conductivity of nanofluids based on oxide and water.

The low temperature dependence of the nanofluid's thermal conductivity indicates that Brownian motion has no effect on the thermal conductivity enhancement at low temperatures. Thermal conductivity improvements between two and four times higher were predicted by the hypothesis in a temperature range of 20 to 50°C[9–12].

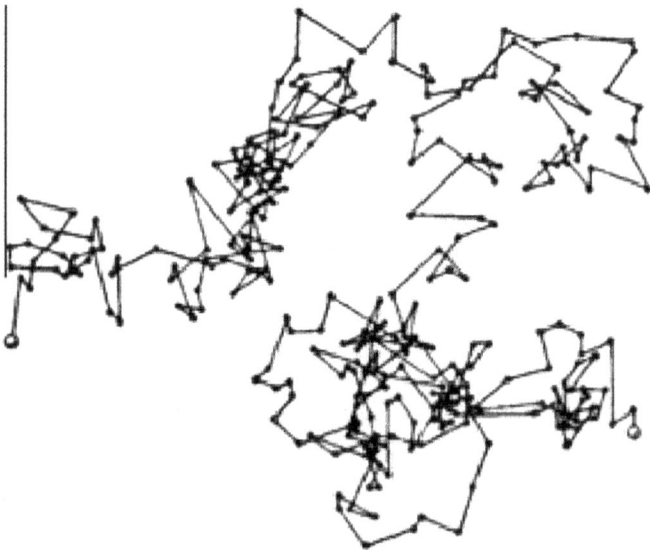

FIGURE 13.1   Brownian motion of a particle suspended in water[1].

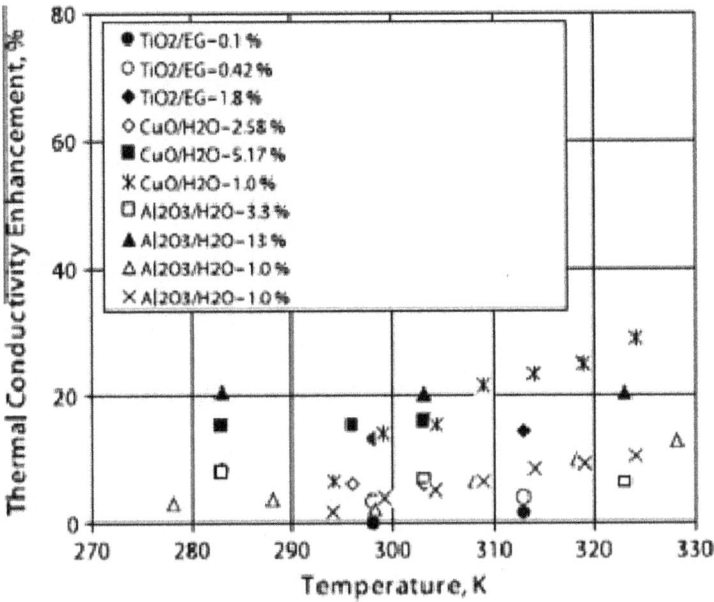

FIGURE 13.2 Effect of the temperature in the enhancement of the thermal conductivity[2].

In the second type of modeling, the nanofluid structure is compared to the structure of a composite material. NPs would make up a central core of this composite material, which would be surrounded by a layer of intermediate-property material, as well as a matrix of base fluid. A multiphase system might be formed by combining these zones, with phase superposition serving as the primary driver of the mixture's increased thermal conductivity.

The liquid molecules at the interface between the solid particles and the base fluid could be much more organized than along the rest of the base fluid. These nanofluids can be affected directly by the thermal behavior of crystallized solids at their interface, which is more conductive than that of liquids.

Figure 13.3 depicts the rise in the interface layer's thermal conductivity as a function of the layer's thickness. Dimensionless increase in thermal conductivity can be represented by the sign j (heat conductivity).

A nanofluid can, however, suffer from a decrease in heat conductivity if an interface layer is formed. Because nanofluids have a high heat resistance, it is not recommended to employ them when the flow's characteristic length is too short or the NPs themselves are too small[13–18].

Thermally equivalent particle clusters are proposed for use in this mechanism's heat transfer. This model's results are in good accord with the experimental data. Excessive particle clustering in nanofluids has recently been observed to reduce thermal conductivity, hence an optimal level of clustering must be determined in order

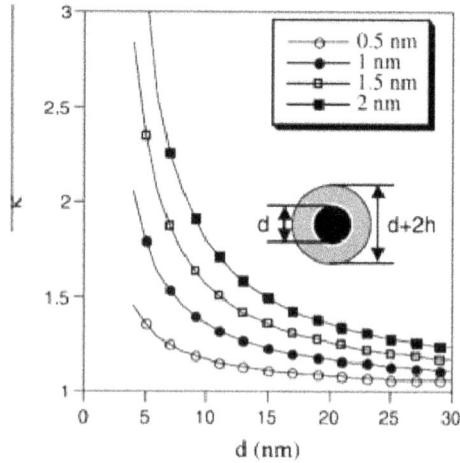

FIGURE 13.3 Effect of the thickness of the interface layer on the thermal conductivity enhancement[2].

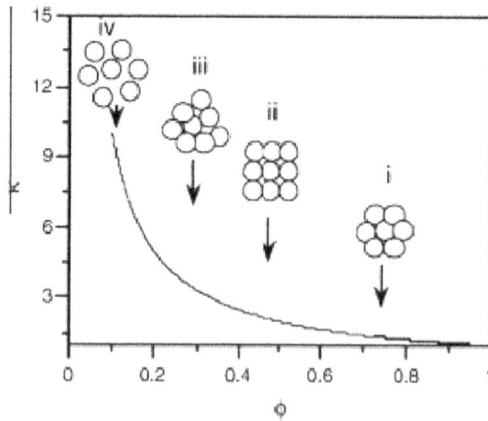

FIGURE 13.4 Effect of the clustering of particles on the enhancement of thermal conductivity[2].

to maximize the effect of particle clustering on thermal conductivity. Figure 13.4 illustrates that particle clusters with less compact shapes improve heat transfer more than compact groupings.

The four primary mechanisms responsible for the increase in thermal conductivity achieved using nanofluids, as described in the models, are as follows[19–22]:

1. NPs' Brownian motion.
2. The formation of layers at the NP-to-base fluid interface.
3. Clustering of particles.
4. Temperature transport in a "ballistic" manner of NPs.

Considering that the thermal diffusion concept is flawed, the study of NPs' ballistic nature is important. Justification for this claim is that the NPs are so small, the heat transmission processes considered in the thermal diffusion model change from a random spread to ballistic behavior when applied to them. Since the ballistic phonon transport mechanisms in a nanofluid increase with particle size, any macroscopic approach to a nanofluid's properties will be limited by this restriction.

It is also possible to increase thermal conductivity by altering pH levels and using transient conduction mechanics in other studies. The Brownian motion-induced effects of nanoconvection and effects related to near-field radiation are mentioned in more recent models.

Thermal conductivity also has a larger impact on convective heat transfer than expected. Since nanofluids flow in both laminar and turbulent circumstances, an explanation for the growth in this coefficient is also being sought. Flow patterns and dimensions, the shape of NPs, and their volumetric fractions along the course of the nanofluid all have an impact on the thermal dispersion coefficient associated with this dispersion.

This coefficient is also a result of the slip between NPs and the base fluid and the disturbances in the flow created by the NPs due to the chaotic movement of particles in the base fluid. Thermal dispersion introduces minor temperature and fluid velocity changes along the flow, which directly alter the conservation equations. The thermal dispersion of nanofluids is studied, and the coefficient of thermal dispersion is correlated with various conditions.

An explanation for the rise in the heat transfer coefficient can be found in a particle migration model. The shear pressures, viscosity gradients, and Brownian motion induced by particle concentration gradients in nanofluids would all contribute to this migration, which would then increase the heat transfer. The results also show that there is an optimal size for nanoparticles in terms of heat transfer and pressure drop.

Analyses of nanofluids' heat transfer coefficient determine the most important variables. Brownian diffusion and thermophoresis were two of the most important factors influencing laminar flows or the viscous layer of turbulent flows. Brownian diffusion increases heat transfer by simply diffusing the heat associated with the particles. Because of the temperature difference that forms surrounding the NPs, the Soret effect, also known as thermophoresis, causes movement.

In contrast, no single mechanism has been able to explain nanofluids. Despite this, the Brownian motion and interface layer theories are substantial advancements that accurately depict observed phenomena[23, 24].

REFERENCES

[1] http://encyclopedia2.thefreedictionary.com.
[2] Pinto, R. V. & Fiorelli, F. A. S., 'Review of the mechanisms responsible for heat transfer enhancement using nanofluids', *Applied Thermal Engineering*, vol. 108, pp. 720–739, 2016.
[3] Das, S. K., Choi, S. U. S., Yu, W. & Pradeep, T., *Nanofluids: Science and Technology*. John Wiley & Sons, 2007.

[4] Saidur, R., Leong, K. Y. & Mohammad, H. A., 'A review on applications and challenges of nanofluids', *Renew. Sustain. Energy Rev.*, vol. 15, pp. 1646–1668, 2011.

[5] Kakaç, S. & Pramuanjaroenkij, A., 'Review of convective heat transfer enhancement with nanofluids', *Int. J. Heat Mass Transf.*, vol. 52, pp. 3187–3196, 2009.

[6] Wang, X. Q. & Mujumdar, A. S., 'A review on nanofluids—Part I: Theoretical and numerical investigations', *Braz. J. Chem. Eng.*, vol. 25, pp. 613–630, 2008.

[7] Wang, X. Q. & Mujumdar, A. S., 'A review on nanofluids—Part II: Experiments and applications', *Braz. J. Chem. Eng.*, vol. 25, pp. 613–628, 2008.

[8] Eastman, J. A., Choi, U. S., Thompson, L. J. & Lee, S., 'Enhanced thermal conductivity through the development of nanofluids', *Mater Res. Soc. Symp. Proc.*, vol. 457, pp. 3–11, 1996.

[9] Liu, M. S., Lin, M. C.-C., Huang, I.-T.& Wang, C.-C., 'Enhancement of thermal conductivity with CuO for nanofluids', *Chem. Eng. Technol.*, vol. 29, no. 1, pp. 72–77, 2006.

[10] Hwang, Y., Par, H. S. K., Lee, J. K. & Jung, W. H., 'Thermal conductivity and lubrication characteristics of nanofluids', *Curr. Appl. Phys.*, no. 6S1, pp. 67–71, 2006.

[11] Yu, W., Xie, H., Chen, L. & Li, Y., 'Investigation of thermal conductivity and viscosity of ethylene glycol based ZnOnanofluid', *Thermochim. Acta*, vol. 491, nos. 1–2, pp. 92–96, 2009.

[12] Mintsa, H. A., Roy, G., Nguyen, C. T. & Doucet, D., 'New temperature dependent thermal conductivity data for water-based nanofluids', *Int. J. Therm. Sci.*, vol. 48, no. 2, pp. 363–371, 2009.

[13] Murshed, S. M. S., Leong, K. C. & Yang, C., 'Enhanced thermal conductivity of TiO2—water based nanofluids', *Int. J. Therm. Sci.*, vol. 44, no. 4, pp. 367–373, 2005.

[14] Patel, H. E., Das, S. K., Sundararajan, T., Nair, A. S., George, B. & Pradeep, T., 'Thermal conductivities of naked and monolayer protected metal nanoparticle based nanofluids: Manifestation of anomalous enhancement and chemical effects', *Appl. Phys. Lett.*, vol. 83, pp. 2931–2933, 2003.

[15] Masuda, H., Ebata, A., Teramae, K. & Hishinuma, N., 'Alteration of thermal conductivity and viscosity of liquid by dispersing ultra-fine particles (Dispersion of g-Al_2O_3, SiO_2, and TiO_2 ultra-fine particles)', *NetsuBussei*, vol. 7, pp. 227–233, 1993.

[16] Lee, S., Choi, S. U. S., Li, S. & Eastman, J. A., 'Measuring thermal conductivity of fluids containing oxide nanoparticles', *ASME J. Heat Transf.*, vol. 121, pp. 280–289, 1999.

[17] Xuan, Y. & Li, Q., 'Heat transfer enhancement of nanofluids', *Int. J. Heat Fluid Flow*, vol. 21, pp. 58–64, 2000.

[18] Xuan, Y. & Roetzel, W., 'Conceptions for heat transfer correlation of nanofluids', *Int. J. Heat Mass Transf.*, vol. 43, pp. 3701–3707, 2000.

[19] Heris, S. Z., Esfahany, M. N. & Etemad, S. G., 'Experimental investigation of convective heat transfer of Al2O3/water nanofluid in circular tube', *Int. J. Heat Fluid Flow*, vol. 28, no. 2, pp. 203–210, 2007.

[20] Kim, D., Kwon, Y., Cho, Y., Li, C., Cheong, S. & Hwang, Y., 'Convective heat transfer characteristics of nanofluids under laminar and turbulent flow conditions', *Curr. Appl. Phys.*, vol. 9, no. 2—Suppl. 1, pp. 119–123, 2009.

[21] Jung, J.-Y., Oh, H.-S.& Kwak, H.-Y., 'Forced convective heat transfer of nanofluids in microchannels', *Int. J. Heat Mass Transf.*, vol. 52, nos. 1–2, pp. 466–472, 2009.

[22] Sharma, K. V., Sundar, L. S. & Sarma, P. K., 'Estimation of heat transfer coefficient and friction factor in the transition flow with low volume concentration of Al_2O_3 nanofluid flowing in a circular tube and with twisted tape insert', *Int. Commun. Heat Mass Transf.*, vol. 36, no. 5, pp. 503–507, 2009.

[23] Wen, D. & Ding, Y., 'Experimental investigation into convective heat transfer of nanofluids at the entrance region under laminar flow conditions', *Int. J. Heat Mass Transf.*, vol. 47, no. 24, pp. 5181–5188, 2004.

[24] Lee, S. & Choi, S. U. S., 'Application of metallic nanoparticle suspensions in advanced cooling systems', in: *International Mechanical Engineering Congress and Exhibition*, Atlanta, GA, pp. 17–22, Nov. 1996.

14 Hybrid Nanofluids

CONTENTS

14.1 INTRODUCTION

In present heat transfer applications, the poor thermal conductivity of conventional fluids makes them unsuitable. The most commonly utilized base fluids are water, oil, and ethylene glycol, all of which have a reduced heat transfer capability due to their lower thermal conductivity. With the addition of particularly thermally conductive nano-sized particles to the base fluid, conventional fluids' heat transfer capacity is improved. The thermophysical and image properties of liquids, such as thermal conductivity, viscosity, oxidative stability, and specific heat capacity, have been dramatically improved by the addition of nanoparticles. Nanofluids can be used in a wide range of industrial applications, including metal cutting, heat exchangers, engine cooling, solar collectors, nuclear cooling, and electronics cooling. Dispersion of nanometer-sized particles into the base fluid is known as "nanofluid" by Choi and Eastman[1–3].

In recent years, research has focused on nanofluids, including the development, characterization, evaluation, and usage of diverse NP combinations in specific base fluids. It is possible to enhance the properties of nanofluids by using various types of NPs, such as metals (Cu and Al and Fe and Al and Ni), metal oxides, metal nitride, metal carbide, and carbon materials (CNTs, SWCNTs, MWCNTs, graphene, graphite, graphene oxide) to improve thermal conductivity, heat transfer, and nanofluid stability. These metals and metal oxides were created in a one- and two-step process. One-step manufacture of metallic nanofluids is the most efficient method since it

DOI: 10.1201/9781003163633-14

minimizes the formation of agglomeration. Nanofluids including metal oxide and carbon nanomaterials can be developed using the two-step process, which is a systematic approach. In order to obtain the desired stability and prevent agglomeration, this approach employs ultrasonication, high shear homogenization, surfactants, and pH modification. Several factors, such as type, size, shape, concentration, the stability of suspended NPs, base fluid type, and fluid temperature all have an impact on the thermal conductivity of nanofluid NPs[4, 5].

NP suspensions of Al_2O_3 (10–20 nm) and SiO_2 (40–50 nm) were tested for thermal conductivity in pure methanol at 293.15°C. Al_2O_3 and SiO_2 showed a 10.74% and 14.29% increase above base fluid at 0.5 vol%, respectively. SiC NPs dispersed in nanofluids with water content of 0–1 vol% have been observed. An inverse relationship between conductivity and particle size was shown to be noticeable in the nanofluid at 0.1 vol% of NPs, according to this study. In order to determine the heat conductivity of a fluid, KD2-Pro thermal analyzers are utilized. However, single NP dispersion in the base fluids has a number of limitations that need to be considered. The most critical issue was the thermal system's inability to operate properly due to nanofluids being clogged in the flow channels as they transit through the device. Because NPs damage the flow channel walls, their performance is also affected by this corrosive effect. Different conventional fluids are mixed with hybrid NPs to create these hybrid nanofluids, which can then be used in a variety of ways. Because of their exceptional thermal properties, hybrid nanofluids have attracted a lot of attention as a thermal/working fluid. Water, ethylene glycol, oil mixes, and ethylene/water combinations are the most commonly employed fluids for the synthesis of hybrid nanofluids. Nanofluids are stable when the particle size is kept below 100 nm, which is crucial for the synthesis of hybrid nanoparticles. Nanofluids as a heat transfer fluid, however, are only in the early stages of research based on simulation, and at present, people are focusing on understanding deeply the changes introduced in thermophysical properties and improving fluid heat transfer characteristics after the inclusion of nanoparticles. These synergistic nanofluids are employed in cooling and heating systems. The absorber plate's heat is immediately transferred to the circulating fluid, increasing the PV/T system's efficiency and performance. For cooling photovoltaic panels, water or air circulates to increase the efficiency of the solar cell. Due to their better thermal behavior, nanofluids are increasingly used to substitute working fluids in PV cells to further improve the cooling effect. With their higher thermal conductivity, nanofluids can more effectively replace base fluids and increase system performance[6–11].

Classification of nanofluids can be done in a variety of ways, including by the type of constituent material, the basic fluid properties, or the specific NP qualities. This is mostly due to the fact that it is possible to track the chronological progression of nanofluids in this manner.

14.2 EXPERIMENTAL STUDIES OF SHC OF HYBRID NANOFLUIDS

A miniature heat exchanger was used to test the thermal performance of an Al_2O_3-CuO hybrid based on ethylene glycol. In comparison to Al_2O_3 and CuO mono NPs, the performance of the hybrid nanofluid is being studied. Volume concentrations of 0.02%, 0.04%, and 0.06% of the nanofluid were generated at a 60:40 mixing ratio. It was found that the optimum SHC value was 2558KJ/kg K when the volume concentration was 0.02 percent and fell by 0.82% and 2.18% when it was 0.04% and 0.06% at a constant temperature of 30°C. Hybrid nanofluids, on the other hand, demonstrated a reduced SHC compared to mono nanofluids, except at a volume concentration of 0.06%. The experimental results of the SHC of the nanofluids investigated show that the hybrid nanofluids will gain and dissipate heat more quickly, and hence present a better fluid for heat transfer purposes. A similar study was done on the use of hybrid nanofluids in a micro channel heat exchanger. At a volume concentration of 0.1%, Al_2O_3-Graphene hybrid nanofluids are created. SHC of hybrid nanofluids was shown to be lower than the basic fluid, according to their findings. When compared to graphene and alumina nanofluids, the SHC was found to be lower than alumina, but higher than graphene in terms of performance[12].

14.3   HYBRIDIZATION EFFECT ON SHC OF HYBRID NANOFLUIDS

There is a decrease in the nanofluid's heat capacity when compared to that of the base liquid, as seen in Figure 14.1. Nanofluid specific heat is lower when the specific

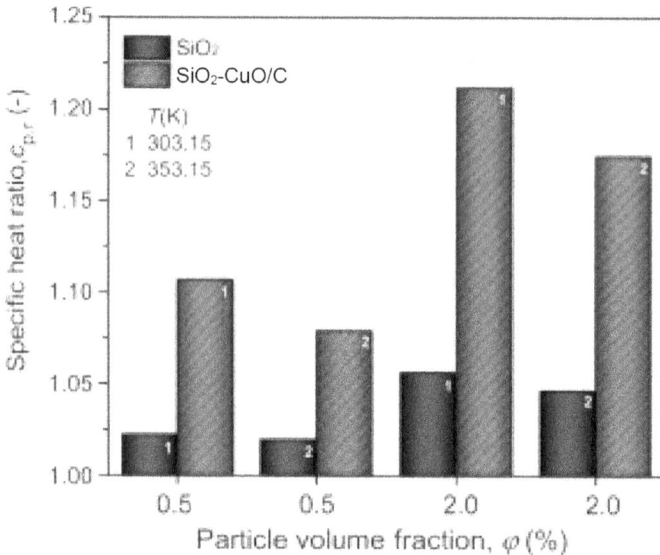

FIGURE 14.1   Effect of hybridization on specific heat ratios across temperature and volume fraction[12].

heat ratio is high. The SHC of the hybrid nanofluid is lower than the SHC of the SiO_2 nanofluid at all volume concentrations and temperatures investigated. It has been found that a 5.7% SHC reduction is optimal. Hybrid nanofluids, however, have a significant impact on SHC reduction due to factors such as particle size and volume concentration.

14.4 EFFECT OF TEMPERATURE ON SHC OF HYBRID NANOFLUIDS

The SHC of hybrid nanofluids can be significantly affected by temperature. Everyone in the field is in agreement that hybrid nanofluids have a lower SHC than water. As the thermal diffusivity of hybrid nanofluids increases, so does their SHC behavior. SHC's response to temperature varies widely, making it difficult to draw any firm conclusions. The SHC of the MWCNT-CuO decreased from 20 to 35°C in their investigation (see Figure 14.2a). Temperatures over 35°C were shown to have the greatest SHC reduction of 2.13%. However, over 35°C, the SHC increased (see Figure 14.2c). While the SHC of a CNT-Al_2O_3 showed a growing tendency between 20° and 65°C and subsequently reduced until 80°C, after which it surged again. While the CNT-Al_2O_3 synthesized at 0.1% volume concentration showed a steady increase in SHC with temperature, the hybrid CNT-Al_2O_3 showed a steady decrease in SHC. CNT-SiO_2 was synthesized at a temperature ranging from 25 to 38°C.

At all volume concentrations produced, the SHC was found to be decreasing. Lower volume concentrations resulted in a greater drop in volume (see Figure 14.2a).

In solar collectors, it was found that the thermophysical properties of hybrid nanofluids were more enhanced than for the basic heat transfer fluids, Figure 14.3. Nusselt number increases significantly when hybrid nanofluids, particularly the Cu-MgO hybrid, are used at a volume concentration of 0.02%. Water as a base fluid with hybrid nanoparticles produced outcomes that were less than optimal in terms of both increasing thermal conductivity and reducing pressure drop. It was also shown those NPs in 60 EG:40 W resulted in a significant increase in viscosity, which led to a significant decrease in pressure penalty as well. It was found that the 0.02 Ag-MgO/H_2O nanoliquids had the highest collector efficiency when the Re was increased.

It has been revealed, however, that increasing the concentration of Al_2O_3-Cu nanoparticles results in decreased heat transfer because the dynamic viscosity increases, as shown in Figure 14.4. An Al_2O_3/Syltherm 800 and TiO_2/Syltherm 800 parabolic trough solar collector LS-2 PTC module's results and thermal enhancement were compared, as were the results of using hybrid nanofluids (Al_2O_3 and TiO_2 distributed on Syltherm 800 as the host working fluid). Nanofluids with mono and hybrid properties were compared at inlet temperatures between 300 K and 700 K[13].

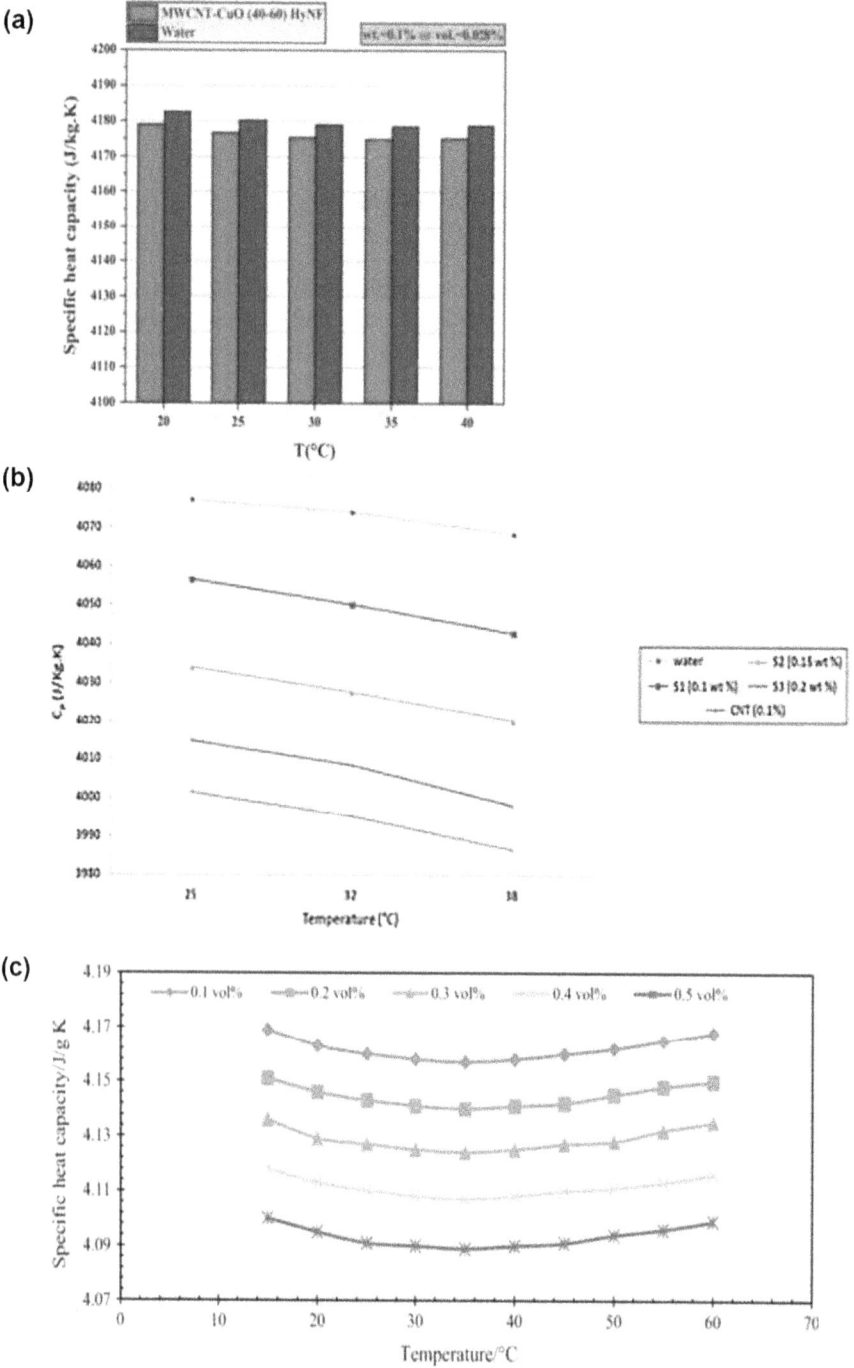

FIGURE 14.2 Decreasing SHC with increase in temperature (a), (b), (c)[12].

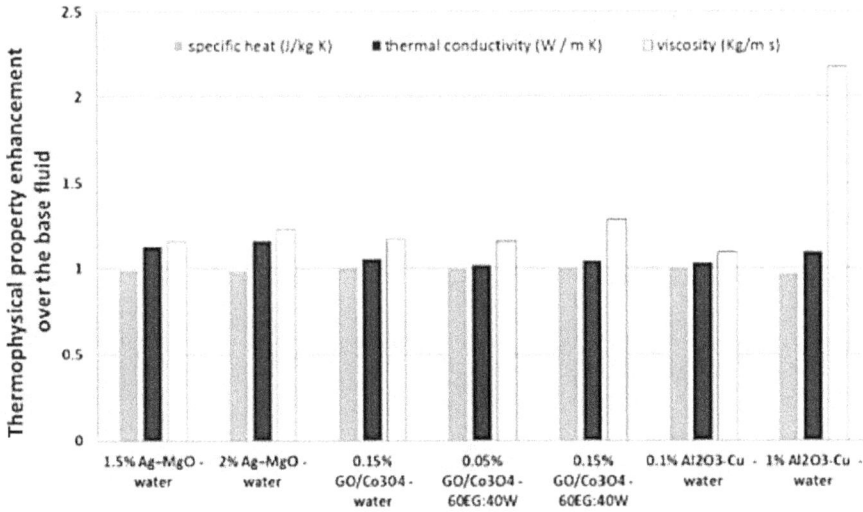

FIGURE 14.3   Thermophysical properties improvement of the used hybrid nanofluids[13].

14.5   APPLICATIONS OF HYBRID NANOFLUIDS

The stability of hybrid nanofluids is still a long-term mystery that has to be solved, which could have a profound impact on their practical uses. It is the most sophisticated engineered fluid that increases the thermal conductivity of base fluids with the dispersion of diverse functional NPs, resulting in superior thermophysical properties that can be applied to a wide range of applications.

The lack of dispersion stability of hybrid nanofluids has been considered a long-term concern in the research community, which has stifled their development and practical applications. Stability can be improved by sonication, mechanical stirring, surface modification, addition of surfactants, and other approaches to address this issue; nevertheless the dispersion issues and mechanisms that are studied by investigating the forces that are deployed between them are still contested[14].

14.5.1   DISPERSION CHALLENGE AND MECHANISM

After the synthesis of hybrid nanofluids, a number of characteristics, problems, and methods must be taken into account to ensure their long-term viability. The most difficult part is to detect microscopic forces, inter-particle forces, other surface modification agents, and external variables that alter the kinetics and dispersion behavior, resulting in the development of NP aggregation in the base fluid (Figure 14.5a–c).

The base fluid nanoparticle normally favors strong van der Waals forces of attraction; they bind and form clumps due to size increment and density, a difference in gravity forces NPs to settle down.

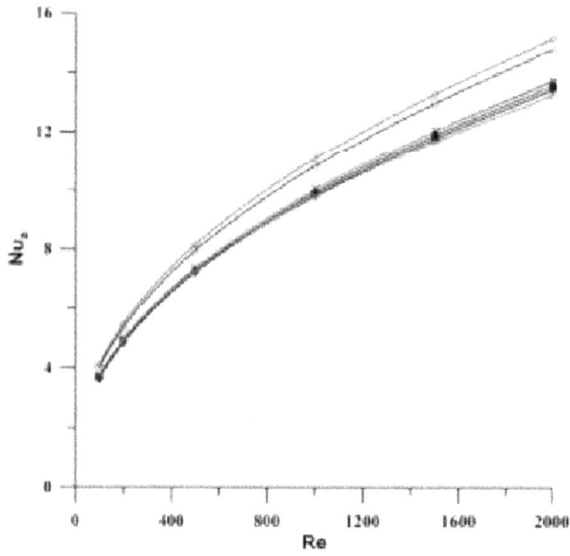

(a)

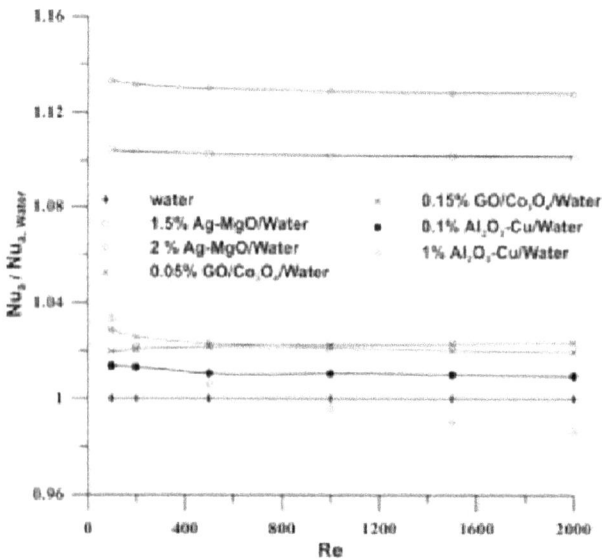

(b)

FIGURE 14.4   (a) Mean Nusselt number (Nua) and (b) ratio of Mean Nusselt number to the
pure water Mean Nusselt number (Nua/Nua-water) versus Reynolds number
for different water-based hybrid nanoliquids[13].

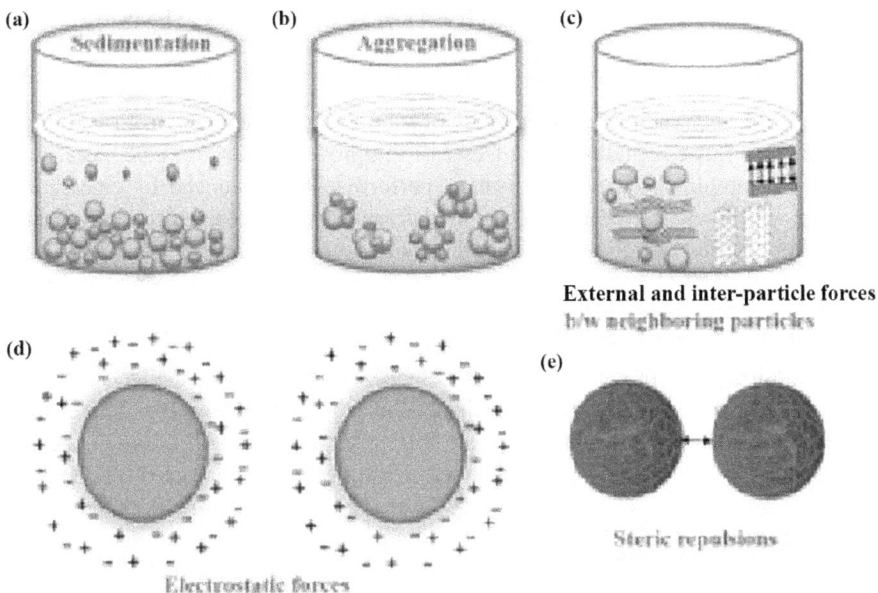

FIGURE 14.5 Schematic representation of nanofluid dispersion: (a) density difference between NPs and base fluid, the gravity force tends to sedimentation of NPs; (b) various forces acting on the NPs tends to aggregation; (c) forces acting on NPs, interactions between neighboring NPs and inter-particle distance between nanorods and sheets; (d) electrostatic forces; and (e) steric repulsions[14–16].

Inter-particle force acting on the particle has also played an important role in controlling nanofluid dispersion. Aside from the surface area, larger nanomaterials such as rod-shaped sheets have a greater affinity for aggregation formation than smaller sphere-shaped NPs. Before aggregation, gravitational forces have less of an effect on dispersion behavior than Brownian motion.

The dispersion process is critical to the stability of NPs, as is the intricacy of the mechanism itself. When NPs with the same surface charges, either positive or negative, come into contact with each other, they will experience difficulties in dispersing and stabilizing them (Figure 14.5d). When it comes to NPs, neutralizing the total charge around them is a primary goal of this technique.

Adsorption, substitution of ions, dissociation of charges at the surface, and depletion of electrons at the surface can all be used to meet this need, but the question is how effectively it reduces the development of aggregation. Furthermore, it is difficult to disperse the aggregates after they have formed. Colloidal suspensions use steric stabilization (Figure 14.5e) as their primary dispersion mechanism. In water-based nanofluids, this stabilization process has a limited temperature range since water freezes and vaporizes. While capping agents or polymer chains play a significant role in oil-based nanofluids, the synthesis of stable nanoparticles' incompatibility with the base fluid plays a vital function in the suspension of stable nanoparticles[14–16].

14.6   IMPACT OF HYBRID NANOFLUIDS IN PV/T SYSTEMS

14.6.1   Exergy

Exergy in thermodynamics is the amount of useful work done by engineering analysis of energy flows in industrial processes to improve the design and reduce total energy consumption. The PV/T system's performance is balanced by calculating the exergy efficiency by considering the characteristics of different energy forms. Optical fibers, concentrators, and nanofluids are preferred due to the exergetic performance of the PV/T system. It has been shown that the exergy efficiency of different PV/T systems can be affected by a variety of factors, including solar radiations, ambient temperature, packing factor; collector length; the nanofluid concentration. Exergy destruction from technical improvements to a system using solar air collectors is being recognized with a new enhanced exergy approach. Exergy and energy efficiency in PV/T systems have received little research attention. Using an aluminum plate and straight and helical channel hybrid PV/T system, 11.1–13.5% exergy efficiency and 59.3–92% energy efficiency were discovered for the PV/T system. It has also been found that PV efficiency has increased by 17.7–38% and thermal efficiency has increased by 31.6–59%. Exergy efficiency in the range of 15% for PV/T, 55% thermal efficiency, 10–13.7% PV efficiency and 90% PV/T energy efficiency were all found[17–19].

14.6.2   Entropy

Any thermodynamic system's performance can be summed up by its thermodynamic performance alone. Several researches have supported the irreversible study of various systems and demonstrated that the formation of entropy is a valuable instrument to select the development efficiency. Due to friction, mixing, chemical reaction, and heat transfer over a finite temperature differential, the generation of entropy is measured[20].

Entropy creation has been the subject of various studies, and some of them have been highlighted here:

1. The usage of nanofluids, NP volume concentration, and channel size and type of flow distribution all help to reduce entropy formation, which in turn support system optimization.
2. The thermophysical properties of nanofluids may conflict with the forecast for entropy formation due to a number of different models.
3. Additions of NPs may increase entropy formation in microchannel studies.
4. When using nanofluids in a flow system, it is often considered that the supply of heat transfer dominates entropy creation, resulting in a substantial decrease in entropy generation due to a uniform temperature.

The rate of entropy generation theory has been extensively researched and documented, and it may be used to compare the performance of various systems[21].

14.6.3 PUMPING POWER

In comparison to base fluids, nanofluids have a higher probability of improving heat transfer coefficient and thermal conductivity. Because of the potential for increased viscosity due to NP dispersion in the base fluid, significant pumping power is required to keep the fluid within the system. High pumping power is a result of increased viscosity, which increases the friction factor. Pumping power is determined by the amount of heat transported and the concentration of particles. Furthermore, the pressure drop is influenced by friction, and pumping power is directly proportional to pressure drop. Because of density gradients and viscous forces, an increase in the nanocomposite volume concentration leads to an increase in the hybrid nanofluid friction factor and pressure drop. The viscosity of hybrid nanofluids is increased by the creation of nano-agglomerates and the re-framing of hybrid nanocomposites with primary NPs. The use of pin fins to improve heat transfer results in better overall performance when increased pressure losses are taken into account through a fin array. The usage of cylindrical pin fins in the inline and staggered heat transfer topologies is reported. When using pin fins, the heat transfer coefficient was found to be 100% more than when using the end wall. This $Cu-TiO_2$/water has been made with varying fractions of NP volume and experimentally evaluated to see how friction factor and pressure drop change. When compared to base fluids and mono fluids, hybrid nanofluids have a large pressure drop. Wall shear rate increases as volume fraction increases because of the inclusion of hybrid NPs[22–24].

14.7   EFFECT OF THE INCLINATION ANGLE

The quantity of heat transferred from hybrid nanofluid to water coolant depends on the temperature differential between the inlet and output fluid flow, the specific heat, and the total mass flow rate of nanofluid considered.

The influence of inclination angle depends on a variety of other factors, such as:

1. Heat transfer rate.
2. The evaporation heat transfer coefficient.
3. The condensation heat transfer coefficient.
4. The total thermal resistance of two-phase closed thermophyson (TPCT).

In addition to the inclination angle, other factors such as tube length, fluid density, acceleration from gravity force, and the inclination angle of the TPCT to the horizontal axis might affect the transfer rate of liquids. The evaporation heat transfer coefficient of nanofluids or water is dependent on the concentration of NPs and the inclination angle (maximum evaporation at 60°C).

The condensation heat transfer coefficient for nanofluids is maximum at lower working temperatures and less at higher operating temperatures, depending on the rate of heat transfer, the concentration of NPs, and the inclination angle (maximum condensation at 30°C). With increasing NP concentration and angle of incidence, TPCT's overall thermal conductivity is reduced[25–29].

14.7.1 HEAT TRANSFER

Particles in the nanofluid collide with one other at the nanoscale level to exchange heat flow between the solid and liquid interfaces between nanofluid and the system surface. The heat exchanger's heat transfer area is the most important factor in determining the thermal system's overall size. It is also important to consider geometric aspects such as tube diameters, pitch, and tube alignment when it comes to influencing the heat transfer coefficient of the heat exchanger. In addition, other variables, such as heat capacity, thermal conductivity, viscosity, density, and surface tension if a phase change is occurring, must be taken into consideration[30].

14.8   CRITICAL CHALLENGES IN USING HYBRID NANOFLUIDS

14.8.1 CHALLENGES OF HYBRID NANOFLUIDS

The effects of the hybrid nanofluids are discussed in the preceding sections, which can yield outstanding outcomes. However, there are still some issues that need to be taken into account:

In the end, the system's stability is the most difficult component, and it may transform even the best results into a nightmare. Performance and heat conductivity are mostly affected by cluster formation. The primary reason for investigating pH stability is the impact on pumping power by a variety of NP combinations, base fluid types, and mixer techniques such as magnetic stirrer and ultrasonicator.

Furthermore, surface modification and incompatibility with the base fluid must be taken into consideration. NP heat transport would be restricted by the stabilizing action and changed functional/oxygen functional groups on the surface/plane. Changing the surface chemistry of the NPs in a combination can lead to an increase in NP suspension in a fluid. Surfactants can also alter the surface properties of nanofluids, resulting in a decrease in surface tension, which can have a negative impact on the nanofluid's thermal and physical properties. Adding excessively many additives to nanofluids can have a negative effect on the stability, viscosity, and thermal conductivity of the fluids.

Recent developments in heat transfer devices have made use of hybrid nanofluids, thanks to their improved thermal characteristics. A single NP suspended in fluid has a lower pressure drop and pumping impact than a suspension of two separate individual nanomaterials, which may result in an increase in viscosity. When it comes to making nanofluids, one of the biggest challenges is figuring out how to keep the manufacturing costs down while still producing high-quality materials. An innovative and expensive apparatus is required to carry out the synthesis of hybrid nanofluids. Nanofluids made with the one-step process have a high stability, but researchers still face difficulties such as high equipment costs, a low production rate, and the high cost of raw ingredients.

In the case of the two-step procedure, the commercial production of nanofluids in bulk quantity is prohibiting the synthesis of NPs. A NP-like magnetic stirrer and ultrasonication and a probe sonicator, however, are the principal obstacles to this technique because of their high costs[31–39].

The physics of heat transport in nanofluids and the different parameters governing the behavior of nanofluids must also be thoroughly understood. A new type of nanofluid, and hybrid nanofluids, are still in the early stages of research and development in the industrial sector. The usage of various hybrid nanofluids for the same application has been shown to produce better results. The thermophysical features of hybrid nanofluids, however, limit their usage in research to a small number of applications. Hybrid nanofluids have been shown to be less effective than single nanofluids, which may be owing to the absence of synergistic effect or correct bonding. Solar thermal collectors, micro-power generators, grinding, heat exchangers, solar energy, heat sinks are some of the applications of hybrid nanofluids. They have improved over time in heat transmission, and it is expected that their performance will increase in a wide range of applications[40–58].

REFERENCES

[1] Yu, W. & Xie, H., 'A review on nanofluids: Preparation, stability mechanisms, and applications', *J. Nanomater.*, vol. 2012, p. 1, Jan. 2012.

[2] Eastman, J. A., Choi, S. U. S., Li, S., Yu, W. & Thompson, L. J., 'Anomalously increased effective thermal conductivities of ethylene glycol-based nanofluids containing copper nanoparticles', *Appl. Phys. Lett.*, vol. 78, pp. 718–720, Feb. 2001.

[3] Ahuja, A. S., 'Augmentation of heat transport in laminar flow of polystyrene suspensions: I. Experiments and results', *J. Appl. Phys.*, vol. 46, no. 8, pp. 3408–3416, 1975.

[4] Choi, S. U. S. & Estman, J. A., 'Enhancing thermal conductivity of fluids with nanoparticles', *ASME-Publications-Fed*, vol. 231, pp. 99–106, Jan. 1995.

[5] Devendiran, D. K. & Amirtham, V. A., 'A review on preparation, characterization, properties and applications of nanofluids', *Renew. Sustain. Energy Rev.*, vol. 60, pp. 21–40, Jul. 2016.

[6] Das, P.K., Mallik, A. K., Ganguly, R. & Santra, A. K., 'Synthesis and characterization of TiO2-water nanofluids with different surfactants', *Int. Commun. Heat Mass Transf.*, vol. 75, pp. 341–348, Jul. 2016.

[7] Mintsa, H. A., Roy, G., Nguyen, C. T. & Doucet, D., 'New temperature dependent thermal conductivity data for water-based nanofluids', *Int. J. Therm. Sci.*, vol. 48, no. 2, pp. 363–371, 2009.

[8] Pang, C., Jung, J.-Y., Lee, J. W. & Kang, Y. T., 'Thermal conductivity measurement of methanol-based nanofluids with Al2O3 and SiO2 nanoparticles', *Int. J. Heat Mass Transf.*, vol. 55, nos. 21–22, pp. 5597–5602, 2012.

[9] Manna, O., Singh, S. K. & Paul, G., 'Enhanced thermal conductivity of nano-SiC dispersed water based nanofluid', *Bull. Mater. Sci.*, vol. 35, no. 5, pp. 707–712, 2012.

[10] Makki, A., Omer, S. & Sabir, H. 'Advancements in hybrid photovoltaic systems for enhanced solar cells performance', *Renew. Sustain. Energy Rev.*, vol. 41, pp. 658–684, Jan. 2015.

[11] Tyagi, V. V., Kaushik, S. C. & Tyagi, S. K., 'Advancement in solar photovoltaic/thermal (PV/T) hybrid collector technology', *Renew. Sustain. Energy Rev.*, vol. 16, no. 3, pp. 1383–1398, 2012.

[12] Adun, H., Wole-Osho, I., Okonkwo, E. C., Kavaz, D. & Dagbasi, M., 'A critical review of specific heat capacity of hybrid nanofluids for thermal energy applications', *Journal of Molecular Liquids*, vol. 340, p. 116890, 2021.

[13] Xiong, Q., Altnji, S., Tayebi, T., Izadi, M., Hajjar, A., Sundén, B. & Li, L. K., 'A comprehensive review on the application of hybrid nanofluids in solar energy collectors', *Sustainable Energy Technologies and Assessments*, vol. 47, p. 101341, 2021.

[14] Vaka, M., Walvekar, R., Rasheed, A. K., Khalid, M. & Panchal, H., 'A review: Emphasizing the nanofluids use in PV/T systems', *IEEE Access*, vol. 8, pp. 58227–58249, 2019.

[15] Dey, D., Kumar, P. & Samantaray, S., 'A review of nanofluid preparation, stability, and thermo-physical properties', *Heat Transf.-Asian Res.*, vol. 46, no. 8, pp. 1413–1442, 2017.

[16] Yu, F., Chen, Y., Liang, X., Xu, J., Lee, C., Liang, Q., Tao, P. & Deng, T., 'Dispersion stability of thermal nanofluids', *Progr. Natural Sci. Mater. Int.*, vol. 27, no. 5, pp. 531–542, 2017.

[17] Aravind, S. S. J. & Ramaprabhu, S., 'Graphenemultiwalled carbon nanotube-based nanofluids for improved heat dissipation', *Rsc Adv.*, vol. 3, no. 13, pp. 4199–4206, 2013.

[18] Aravind, S. S. J. & Ramaprabhu, S., 'Graphene wrapped multiwalled carbon nanotubes dispersed nanofluids for heat transfer applications', *J. Appl. Phys.*, vol. 112, no. 12, 2012, Art. no. 124304.

[19] Yarmand, H., Gharehkhani, S., Shirazi, S. F. S., Amiri, A., Montazer, E., Arzani, H. K., Sadri, R., Dahari, M. & Kazi, S. N., 'Nanofluid based on activated hybrid of biomass carbon/graphene oxide: Synthesis, thermophysical and electrical properties', *Int. Commun. Heat Mass Transf.*, vol. 72, pp. 10–15, Mar. 2016.

[20] Yarmand, H., Gharehkhani, S., Ahmadi, G., Shirazi, S. F. S., Baradaran, S., Montazer, E., Zubir, M. N. M., Alehashem, M. S., Kazi, S. N. & Dahari, M., 'Graphene nanoplatelets-silver hybrid nanofluids for enhanced heat transfer', *Energy Convers. Manage.*, vol. 100, pp. 419–428, Aug. 2015.

[21] Zubir, M. N. M., Badarudin, A., Kazi, S. N., Huang, N. M., Misran, M., Sadeghinezhad, E., Mehrali, M., Syuhada, N. I. & Gharehkhani, S., 'Experimental investigation on the use of reduced graphene oxide and its hybrid complexes in improving closed conduit turbulent forced convectiveheat transfer', *Exp. Therm. Fluid Sci.*, vol. 66, pp. 290–303, Sep. 2015.

[22] Sundar, L. S., Singh, M. K., Ferro, M. C. & Sousa, A. C. M., 'Experimental investigation of the thermal transport properties of grapheme oxide/Co3O4 hybrid nanofluids', *Int. Commun. Heat Mass Transf.*, vol. 84, pp. 1–10, 2017.

[23] Mohammed, H. A., Al-Aswadi, A. A., Shuaib, N. H. & Saidur, R., 'Convective heat transfer and fluid flow study over a step using nanofluids: A review', *Renew. Sustain. Energy Rev.*, vol. 15, no. 6, pp. 2921–2939, Aug. 2011.

[24] Evans, W., Fish, J. & Keblinski, P., 'Role of Brownian motion hydrodynamics on nanofluid thermal conductivity', *Appl. Phys. Lett.*, vol. 88, no. 9, 2006, Art. no. 093116.

[25] Nie, C., Marlow, W. H. & Hassan, Y. A., 'Discussion of proposed mechanisms of thermal conductivity enhancement in nanofluids', *Int. J. Heat Mass Transf.*, vol. 51, nos. 5–6, pp. 1342–1348, 2008.

[26] Jang, S. P. & Choi, S. U. S., 'Role of Brownian motion in the enhanced thermal conductivity of nanofluids', *Appl. Phys. Lett.*, vol. 84, no. 21, pp. 4316–4318, 2004.

[27] Keblinski, P., Eastman, J. A. & Cahill, D. G., 'Nanofluids for thermal transport', *Mater. Today*, vol. 8, no. 6, pp. 36–44, 2005.

[28] Hwang, Y., Lee, J. K., Lee, C. H., Jung, Y. M., Cheong, S. I., Lee, C. G., Ku, B. C. & Jang, S. P., 'Stability and thermal conductivity characteristics of nanofluids', *Thermochim. Acta*, vol. 455, nos. 1–2, pp. 70–74, 2007.

[29] Keblinski, P., Phillpot, S. R., Choi, S. U. S. & Eastman, J. A., 'Mechanisms of heat flow in suspensions of nano-sized particles (nanofluids)', *Int. J. Heat Mass Transf.*, vol. 45, no. 4, pp. 855–863, Apr. 2001.

[30] Eastman, J. A., Phillpot, S. R., Choi, S. U. S. & Keblinski, P., 'Thermal transport in nanofluids', *Annu. Rev. Mater. Res.*, vol. 34, pp. 219–246, Aug. 2004.

[31] Bang, I. C. & Chang, S. H., 'Boiling heat transfer performance and phenomena of Al2O3-water nano-fluids from a plain surface in a pool', *Int. J. Heat Mass Transf.*, vol. 48, no. 12, pp. 2407–2419, 2005.

[32] Das, S. K., Putra, N. & Roetzel, W., 'Pool boiling characteristics of nano-fluids', *Int. J. Heat Mass Transf.*, vol. 46, no. 5, pp. 851–862, 2003.

[33] Jackson, J. E., *Investigation into the Pool-Boiling Characteristics of Gold Nanofluids*. Columbia, MO: Univ. of Missouri, 2007.

[34] Narayan, G. P., Anoop, K. B. & Das, S. K., 'Mechanism of enhancement/deterioration of boiling heat transfer using stable nanoparticle suspensions over vertical tubes', *J. Appl. Phys.*, vol. 102, no. 7, 2007, Art. no. 074317.

[35] Xue, H. S., Fan, J. R., Hu, Y. C., Hong, R. H. & Cen, K. F., 'The interface effect of carbon nanotube suspension on the thermal performance of a two-phase closed thermosyphon', *J. Appl. Phys.*, vol. 100, no. 10, 2006, Art. no. 104909.

[36] Baby, T. T. & Ramaprabhu, S., 'Synthesis and nanofluid application of silver nanoparticles decorated graphene', *J. Mater. Chem.*, vol. 21, no. 26, pp. 9702–9709, 2011.

[37] Baby, T. T. & Ramaprabhu, S., 'Experimental investigation of the thermal transport properties of a carbon nanohybrid dispersed nanofluid', *Nanoscale*, vol. 3, no. 5, pp. 2208–2214, 2011.

[38] Baby, T. T. & Sundara, R., 'Synthesis of silver nanoparticle decorated multiwalled carbon nanotubes-graphene mixture and its heat transfer studies in nanofluid', *AIP Adv.*, vol. 3, no. 1, 2013, Art. no. 012111.

[39] Ho, C. J., Huang, J. B., Tsai, P. S. & Yang, Y. M., 'Water-based suspensions of Al_2O_3 nanoparticles and MEPCM particles on convection effectiveness in a circular tube', *Int. J. Therm. Sci.*, vol. 50, no. 5, pp. 736–748, 2011.

[40] Suresh, S., Venkitaraj, K. P., Selvakumar, P. & Chandrasekar, M., 'Effect of Al_2O_3-Cu/water hybrid nanofluid in heat transfer', *Exp. Therm. Fluid Sci.*, vol. 38, pp. 54–60, Apr. 2012.

[41] Suresh, S., Venkitaraj, K. P., Hameed, M. S. & Sarangan, J., 'Turbulent heat transfer and pressure drop characteristics of dilute water based Al_2O_3-Cu hybrid nanofluids', *J. Nanosci. Nanotechnol.*, vol. 14, no. 3, pp. 2563–2572, 2014.

[42] Selvakumar, P. & Suresh, S., 'Use of Al_2O_3-Cu/water hybrid nanofluid in an electronic heat sink', *IEEE Trans. Compon., Packag. Manuf. Technol.*, vol. 2, no. 10, pp. 1600–1607, Oct. 2012.

[43] Sundar, L. S., Singh, M. K. & Sousa, A. C. M., 'Enhanced heat transfer and friction factor of MWCNT-Fe_3O_4/water hybrid nanofluids', *Int. Commun. Heat Mass Transf.*, vol. 52, pp. 73–83, Mar. 2014.

[44] Madhesh, D., Parameshwaran, R. & Kalaiselvam, S., 'Experimental investigation on convective heat transfer and rheological characteristics of Cu-TiO_2 hybrid nanofluids', *Exp. Therm. Fluid Sci.*, vol. 52, pp. 104–115, Jan. 2014.

[45] Moghadassi, A., Ghomi, E. & Parvizian, F., 'A numerical study of water based Al_2O_3 and Al_2O_3-Cu hybrid nanofluid effect on forced convective heat transfer', *Int. J. Therm. Sci.*, vol. 92, pp. 50–57, Jun. 2015.

[46] Takabi, B. & Shokouhmand, H., 'Effects of Al2O3-Cu/water hybridnanofluid on heat transfer and flow characteristics in turbulent regime', *Int. J. Mod. Phys. C*, vol. 26, no. 4, 2015, Art. no. 1550047.

[47] Ahammed, N., Asirvatham, L. G. & Wongwises, S., 'Entropy generation analysis of graphene-alumina hybrid nanofluid in multiport minichannel heat exchanger coupled

with thermoelectric cooler', *Int. J. Heat Mass Transf.*, vol. 103, pp. 1084–1097, Dec. 2016.

[48] Wang, M., Lin, C.-H. & Chen, Q., 'Advanced turbulence models for predicting particle transport in enclosed environments', *Building Environ.*, vol. 47, pp. 40–49, Jan. 2012.

[49] Reasor, Jr., D. A., Clausen, J. R. & Aidun, C. K., 'Coupling the lattice-Boltzmann and spectrin-link methods for the direct numerical simulation of cellular blood flow', *Int. J. Numer. Methods Fluids*, vol. 68, no. 6, pp. 767–781, 2012.

[50] Sajjadi, H., Salmanzadeh, M., Ahmadi, G. & Jafari, S., 'Lattice Boltzmann method and RANS approach for simulation of turbulent flows and particle transport and deposition', *Particuology*, vol. 30, pp. 62–72, Feb. 2017.

[51] Karimipour, A., Nezhad, A. H., D'Orazio, A., Esfe, M. H., Safaei, M. R. & Shirani, E., 'Simulation of copper-water nanofluid in a microchannel in slip flow regime using the lattice Boltzmann method', *Eur. J. Mech.-B/Fluids*, vol. 49, pp. 89–99, Jan./Feb. 2015.

[52] Mital, M., 'Semi-analytical investigation of electronics cooling using developing nano-fluid flow in rectangular microchannels', *Appl. Therm. Eng.*, vol. 52, no. 2, pp. 321–327, 2013.

[53] Mital, M., 'Analytical analysis of heat transfer and pumping power of laminar nanofluid developing flowinmicrochannels', *Appl. Therm. Eng.*, vol. 50, no. 1, pp. 429–436, 2013.

[54] Esfe, M. H., Akbari, M., Toghraie, D. S., Karimiopour, A. & Afrand, M., 'Effect of nanofluid variable properties on mixed convection flow and heat transfer in an inclined two-sided lid-driven cavity with sinusoidal heating on sidewalls', *Heat Transf. Res.*, vol. 45, no. 5, pp. 409–432, 2014.

[55] Esfe, M. H., Arani, A. A. A., Karimiopour, A. & Esforjani, S. S. M., 'Numerical simulation of natural convection around an obstacle placed in an enclosure filled with different types of nanofluids', *Heat Transf. Res.*, vol. 45, no. 3, pp. 279–292, 2014.

[56] Ay, C., Young, C.-W. & Young, C.-F., 'Application of lattice Boltzmann method to the fluid analysis in a rectangular microchannel', *Comput. Math. Appl.*, vol. 64, no. 5, pp. 1065–1083, 2012.

[57] Yang Y.-T. & Lai, F.-H., 'Numerical study of flow and heat transfer characteristics of alumina-water nanofluids in a microchannel using the lattice Boltzmann method', *Int. Commun. Heat Mass Transf.*, vol. 38, no. 5, pp. 607–614, 2011.

[58] Holmes, S., Jouvray, A. & Tucker, P. G., 'An assessment of a range of turbulence models when predicting room ventilation', *Proc. Int. Conf. Proc. Healthy Buildings*, vol. 2, pp. 401–406, 2000.

15 Applications of Nanofluids and PCMs

CONTENTS

15.1 APPLICATIONS OF NANOFLUIDS

The Al_2O_3-H_2O nanofluids were investigated as PCMs for thermal energy storage cooling systems. After mixing Al_2O_3 NPs with water, the results showed that the supercooling degree was reduced and the freezing duration was reduced significantly. The nanofluids were frozen for 0.2 wt% Al_2O_3 NPs for 20.5% less time. A novel type of nanofluid (called PCM) containing small TiO_2 NPs suspended in a solution of saturated $BaCl_2$ was developed. The thermal conductivity of the nanofluids was increased when compared to the base fluid. The supply/storage rate was

DOI: 10.1201/9781003163633-15

enhanced as well as the supply/storage capacity of the nanoliquids, which allowed for improved flow rate[1].

In a study, Cu nanofluids (including paraffin) showed an improvement in the overall thermal conductivity and the heat transfer rate when used for heating systems. There were 30.3% and 28.2% decreases in heating and cooling durations, respectively, while heating and cooling the 1.0wt% of Cu NPs in paraffin. An additional investigation discovered that the carbon nanofibers composite conducts heat transfer in a efficient way. Thermal properties of nanofluids are commonly recommended in order to store and transfer thermal energy in cooling and heating systems. However, CuO-oil had the lowest temperature distribution at both high and low rotational speeds, resulting in the most efficient heat transfers. Particle fraction optimization in nanofluid flows maximized thermal performance under acceptable limitations. The heat transfer rate is believed to be small when there are few particles present, but when there are a significant number of particles, the shear stresses and pumping power requirements are much greater[2].

This competition revealed a trade-off opportunity to maximize heat transfer rate at constant pumping power by selecting the optimum number of particles. The cooling performance of silicon microchannel heat sinks can be improved by utilizing nanofluids. Purified water was used to create the nanofluid, which was then diluted with different proportions of small Cu particles. Thermal conductivity and dispersion impacts of increased thermal conductivity and dispersion in nanofluid coolants have been found to significantly improve performance. Because of their small size and low volume percentage, the presence of NPs in the fluid did not generate any further pressure drop. Microheat sinks with two types of nanofluids and conjugate heat transfer (i.e. CuO nanospheres at low volume concentrations in water and in ethylene glycol).

According to research, Brownian motion has less of an impact on fluid viscosity than it does on thermal conductivity. To improve heat transfer in microchannels, they recommended using high Prandtl number fluids and high thermal conductivity NPs. Particles having a dielectric constant near to that of the base fluid and a wall material that minimizes the attraction between particles and the wall should be selected to decrease particle-particle and particle-wall interactions[3].

15.1.1 Industrial Cooling Applications

Using nanofluids in closed-loop cooling cycles might save the electric power industry in the United States roughly 10–30 trillion Btu per year (the yearly energy consumption of about 50,000–150,000 households). Thermal conductivity and specific heat can both be improved by using phase transition materials as NPs in nanofluids. An indium nanoparticle suspension in polyalphaolefin (melting temperature, 157°C) has been created utilizing a one-step, nanoemulsification process. Thermal conductivity, specific heat, viscosity and temperature influence on the thermophysical characteristics of the fluid was determined experimentally. The fluid's effective specific heat was significantly increased by observing the melting and freezing of indium nanoparticles[4].

15.1.2 Smart Fluids

It has become increasingly clear that the absence of adequate clean power sources and the growing use of battery-powered devices, such as smart phones and laptops, has emphasized the need for technologically advanced energy management in this new age of energy consciousness. It has been shown that nanofluids can perform this function as a smart fluid in some cases. A heat valve made from a specific class of nanofluids can be used to regulate the flow of heat in a device. Both low and high states of the nanofluid can be easily achieved, allowing the heat to be dissipated effectively. More evidence of a robust and responsive operating system will be needed to cross the barrier to heating and cooling technologies[5].

15.1.3 Nuclear Reactors

Nanofluids could be used to improve the performance of any water-cooled nuclear system that is limited in heat removal. The primary coolants of pressurized water reactors (PWRs), backup safety systems, accelerator targets, plasma diverters, and so on are all potential applications[6]. For nuclear power plants with a PWR, critical heat flux (CHF) between fuel rods and water is a limiting factor in steam generation. This occurs when vaporized fuel rod bubbles conduct heat at a lower rate than liquid water. With nanofluids, the alumina-coated fuel rods limit the creation of the vapor layer around the rods and thereby increase the CHF substantially by pushing away newly created bubbles. It is also possible to use nanofluids as a coolant in emergency cooling systems, which could contribute to an increase in power plant safety[6].

The amount of NPs carried away by the boiling vapor is unpredictable when using nanofluids in a power plant system. Another worry is the additional safety procedures required for the disposal of nanofluid. Minimal utilization of a nanofluid coolant is envisaged for boiling-water reactors (BWRs) owing to concerns about nanoparticle carryover to the turbine and condenser. The deposition of nanofluids on a structured surface increased the critical heat flux by a significant amount. If the structure and thickness of the deposition sheet can be regulated, it may be possible to increase CHF with little decrease in heat transfer. The boiling of nanofluids, rather than the NPs themselves, has the potential to create an improved surface in a relatively easy manner. Nuclear power facilities could benefit from the usage of nanofluids in the future[7].

There are several critical information gaps at this moment, including the support of the thermal-hydraulic performance of nanofluids in prototype reactor conditions and the compatibility of the nanofluid chemistry with reactor materials. Another potential use for nanofluids in nuclear systems is to mitigate the effects of catastrophic incidents in which the reactor's core melts and falls to the reactor vessel's lowest level. In the event of an accident, it is preferable to keep the molten fuel inside the vessel by eliminating decay heat from the vessel wall. This process is constrained by the presence of CHF on the vessel's outer surface, but an analysis shows that the introduction of nanofluid can boost nuclear reactors' in-vessel retention capabilities by as much as 40%. Water-cooled nuclear power systems have limited CHF, but nanofluid may considerably improve the CHF of coolant such that there is an economic gain while simultaneously enhancing the safety standard of the power plant[8].

15.1.4 EXTRACTION OF GEOTHERMAL POWER AND OTHER ENERGY SOURCES

Nanofluids can be used to cool pipes that are exposed to temperatures between 500°C and 1000°C when extracting energy from the Earth's outside. In high-friction, high-temperature environments, such as drilling, nanofluids can help keep drilling apparatus and equipment cool. Because of their properties as a "fluid superconductor," nanofluids could be used in a PWR power plant system to generate enormous amounts of work energy by extracting energy from the Earth's core.

Using nanofluids to cool sensors and electronics that can operate at higher temperatures in down hole tools, as well as revolutionary improvements utilizing new methods of rock penetration cooled and lubricated by nanofluids, will lower production costs in the drilling technology sub-area, which is critical to geothermal energy. In high-grade deposits, these advancements will allow for access to deeper, hotter, and more commercially viable temperatures. An order of magnitude (or more) increase in reservoir performance and heat-to-power conversion efficiency could be achieved in the subfield of power conversion technology by improving heat transfer performance for low-temperature nanofluids and developing plant designs for higher resource temperatures up to supercritical water[9].

15.1.5 AUTOMOTIVE APPLICATIONS

Conventional automobile thermal systems (radiators, engines, heating, ventilation, and air-conditioning (HVAC)) use synthetic high-temperature heat transfer fluids that have breakable heat transfer capabilities. Nanofluids with increased thermal conductivity, consequently of NPs addition, might prove useful for these applications[10].

15.1.6 NANOFLUID COOLANT

Manufacturers are required to lower the amount of energy required to overcome road wind resistance in order to enhance vehicle aerodynamic designs and ultimately fuel economy. Aerodynamic drag consumes about 65% of a truck's entire energy output when traveling at high speeds. On the other hand, the huge radiator in front of the engine helps to keep things cool by allowing in as much outside air as possible to pass through it. The use of nanofluids as coolants has allowed radiators to be reduced in size and relocated more easily. Consequently, coolant pumps could be lowered in size and automobile engines could be run at higher temperatures, allowing for more horsepower while adhering to rigorous pollution regulations. High-thermal conductive nanofluids can reduce radiator frontal area by up to 10% when used in radiators, according to research. Up to a 5% reduction in fuel consumption can be achieved by reducing aerodynamic drag. With nanofluid, the friction and wear of machines such as pumping systems and compressors was significantly reduced. This resulted in a

6% fuel savings. Future savings gains could be even bigger if this strategy is implemented successfully.

A device that can simulate the coolant flow in a radiator was created and calibrated to examine if nanofluids degrade the material of radiators. Various nanofluids are now being tested and measured for material loss in typical radiator materials. Weight loss measurements as a function of fluid velocity and impact angle are used to estimate the rate of corrosion of radiator material. In tests, nanofluids made from ethylene and tri-cloro ethylenegycols did not degrade at velocity up to 9 m/s or contact angles between 90° and 30°. Copper nanofluid eroded at a velocity of 9.6 m/s and an impact angle of 90°C. Recession rates of 0.065 miles/year of vehicle operating were computed. Preliminary research shows that copper nanofluid has a higher wear rate than base fluid, presumably due to the oxidation of copper nanoparticles. An increase in alumina nanofluids' wear and friction rate over the base fluid was observed in this study. Cast iron grinding with wet, dry, and minimum quantity lubrication (MQL) wheel wear and tribological properties are tested. It was decided to test the MQL grinding process with that of pure water using water-based alumina and diamond nanofluids. Nanofluids have been found to minimize grinding force, improve surface roughness, and prevent the burning of the workpiece[11].

15.1.7   ELECTRONIC APPLICATIONS

Nanofluids are used to cool microchips in computers and other devices, such as smartphones and tablets. Microfluidic applications are also used in a variety of electronic devices.

15.1.8   COOLING OF MICROCHIPS

Reduced chip size is hampered by the inability to effectively dissipate heat, while nanofluids, with their high thermal conductivity, can be used to cool computer devices. The next generation of computer chips is expected to produce localized heat flux above 10 mW/m^2 and total power over 300 W. As a next-generation cooling device, the nanofluid oscillating heat pipe (OHP) cooling system can remove heat fluxes over 10 mW/m^2 when used in conjunction with thin film evaporation. In order to view the oscillation, the OHP's metal pipe system should be modified to make use of transparent glass or plastic. Even though copper-based OHP systems are often composed of glass or plastic, this change in material has an effect on thermal transfer qualities that alter system performance and hence alter experimental findings.

An OHP system's integrity must be maintained while collecting experimental data. An amorphous silicon imaging device's high-intensity neutron beam made it possible for them to capture dynamic images at the rate of 1/30th second. A nanofluid was used to suspend diamond nanoparticles in water. OHPs and nanofluids aren't new discoveries, but their unique features allow the nanoparticles to be completely suspended in the base liquid, which improves their heat transfer capacities. Due to nanofluids' temperature-dependent thermal conductivity and nonlinear connection between thermal conductivity and concentration, they are high-performance conductors with enhanced CHF. The OHP converts the immense heat of a high-powered device into fluid-kinetic energy, which prevents the liquid and vapor phases from interfering with one another due to their similar motion.

Moving the OHP prevents NPs from settling, which improves the cooling device's performance. The difference in temperature between the evaporator and condenser was reduced from 40.9°C to 24.3°C when 80 W of power was applied. The temperature difference between evaporator and condenser does not continue to increase with increasing heat input, but instead increases with increasing oscillation motion. This phenomenon prevents the nanofluid's effective thermal conductivity from rising indefinitely[11].

15.2   BIOMEDICAL APPLICATIONS

Nanodrug Delivery. MEMS-based DNA sequencers, drug delivery microchips, and nanogels and gold-coated NPs are only some of the examples of nanomedicine applications that have been developed in the past few years. Advanced efforts in constructing integrated micro- or nanodrug delivery systems are aimed at monitoring and managing target-cell responses, understanding biological cell activities, or enabling drug development processes. While conventional medication distribution is characterized by the "high-and-low" phenomenon, microdevices allow precise drug delivery through implanted and transdermal techniques. With each dose, the blood concentration of a medicine provided normally will rise and then fall. For each conventionally delivered medicine, this cycle will be repeated.

Drugs can be delivered over a long period of time using nanodrug delivery (ND) devices. It is therefore possible to maintain the desired medication concentration within the therapeutic window. Temperature control was added to the nanofluid and purge fluid flow to ensure that the medicine was delivered to living cells at an ideal temperature, 37°C. The increased wall heat flux also had a positive effect on the uniformity of medication concentrations. The uniformity of nanodrug concentration is influenced by a variety of factors, including the length of the channel, the diameter of the particles, and the Reynolds numbers of the nanofluid and primary microchannel. Convection-diffusion transport relies on channels longer than a few micrometers and particles smaller than a few micrometers in diameter[12, 13].

15.3   FUTURE SCOPE AND CHALLENGES FOR NANOFLUID APPLICATIONS IN VARIOUS INDUSTRIAL SECTORS

In a thermal system, a high-efficiency heat transfer device is a heat pipe. To transmit heat, it employs a phase transition in the working fluid. Assorted nanofluids with CuO contents ranging from 0.5 to 2.0 wt% are used to test the horizontal microgroove heat pipe's heat transfer performance. Heat transfer coefficient improved by 46% when the CuO content was optimized to 1%, and CHF improved by 30% when the CuO concentration was optimized up to 1.0%. The two-phase thermal performance of thermosiphons was examined by employing aqueous Al_2O_3 and CuOnanofluids.

A higher wettability was observed for nanofluids based on Al_2O_3. The closed-loop thermosiphon thermal performance was hampered by a decrease in active bubble nucleation sites. Electronics development over the last few decades has led to cooling being a significant factor in total system performance. In order to achieve the desired efficiency and reliability, the tremendous heat flux created by the CPU chip must be quickly cooled. Modern electronics' most efficient cooling systems use micro- and mini-channels. When it comes to electronic cooling, nanofluids based on Al_2O_3 and TiO_2 are the most commonly utilized materials[14].

It was found that as the final process temperature declined, the convective heat transfer coefficient grew. A 0.5 vol% concentration of Al_2O_3 particles improved the heat transfer coefficient by 6%. Static and dynamic single-phase models were used to explore the thermal and hydraulic properties of micro-fin heat sinks. Nanofluids containing ZnO were employed. The heat transfer coefficient develops from 4300 to 11,470 W/m^2K as the volume fraction increased from 1.5 to 3.0%. The heat transfer performance of Al_2O_3 (35 nm) and TiO_2 (50 nm) nanofluids in spiral coiled tubes was experimentally examined over a large Reynolds number range (50–4500). Dispersed and suspended in water were nanoparticles with a volume fraction of 0.25–1.0%.

Nanofluids of Al_2O_3 and TiO_2 were stabilized by Triton X-100 and Cetyltrimethylammonium Bromide, respectively. With Al_2O_3 dispersions, the heat transfer performance was improved more than with nanofluids containing TiO_2. At Re = 1865, the nanofluid of 1.0 vol% Al_2O_3 showed the best performance. In addition, the pressure drop is considerably exacerbated when the NP concentration is above 1%. NP-enhanced metal oxide coolants are used in drilling, milling, grinding, and turning. Nanofluid machining outperforms conventional flood cooling and dry machining in terms of surface finish and cooling effectiveness. The effect of Al_2O_3-TRIM E709 emulsifier nanofluid as a cutting lubricant in grinding EN-31 steel was studied. The high specific energy requirements of the grinding process led to elevated temperatures in the grinding zone.

It was found that nano-lubricants reduced the temperature of the wheel working components by 20 to 30%. Using aqueous TiO_2 nano-minimum amount lubrication, the flank wear rates of Al-6061 alloys were examined (MQL). It was compared to conventional oil-based MQL by adding 1.5% volume fraction NPs. Cutting tool edge chipping was reduced when using this lubricant in comparison to traditional oil-based lubricants. This was possible thanks to the cutting tool's edge remaining intact due to the nanofluid's faster cooling rate. Pure Al_2O_3 nanofluids derived from coconut oil have recently been tested for their ability to convert Al-7079 TiC in-situ metal matrix composites. From 0.1 to 0.6 vol% of NPs were used in the experiment[14].

15.3.1 Application of Nanoencapsulated PCMs in Energy Storage

PCMs can be used in a variety of industries to store and manage energy. Heat transfer coefficient can be increased by up to 70% by using NP-Enhanced Phase Change Materials (NEPCM) slurry with a 28% particle volume fraction. NEPCMs slurry has also been studied for its hydrodynamic and thermal properties, and the results showed that adding NEPCMs to the base fluid resulted in a significant increase in heat transfer. While PCM thermal conductivity is critical for solar energy, waste heat recovery, and temperature regulation, their optimum phase transition temperature and cost efficiency are critical considerations in smart buildings. A variety of applications for energy management and storage are discussed in the following sections, including nanoencapsulated PCMs.

15.3.2 THERMAL MANAGEMENT OF ELECTRONICS

Thermal management in electronic equipment has been impacted by the increase in processing power. Laptops and desktop computers can have their heat loads dispersed by using a heat sink and fan. Thus, PCMs can be employed to control the temperature of portable electrical gadgets (such as smart phones). In this case, the released heat melts PCMs and then the PCMs release their heat to the environment. When the PCMs are completely melted, if the electrical devices are still on heat, the temperature will rise through sensible heating (Figure 15.1). Paraffins are frequently used to keep temperatures between 40°C and 45°C.

Paraffin and expanded graphite composites are employed in a battery temperature control strategy. The results showed that the use of a paraffin-coated heat pipe module reduced fan power usage. A volume percentage of Al_2O_3 NPs is mixed with tricosane in order to develop nano-enhanced PCMs as an energy storage medium for electronic cooling applications. Nano-enhanced PCMs have been shown to reduce the evaporator temperature by around 25.75%, which in turn saves roughly 53% of the fan power. PCM nanocapsules coated with silver had an increase in thermal conductivity from 0.240 to 1.346 W/m K (n-octacosane core, silica shell) when exposed to high temperatures[15].

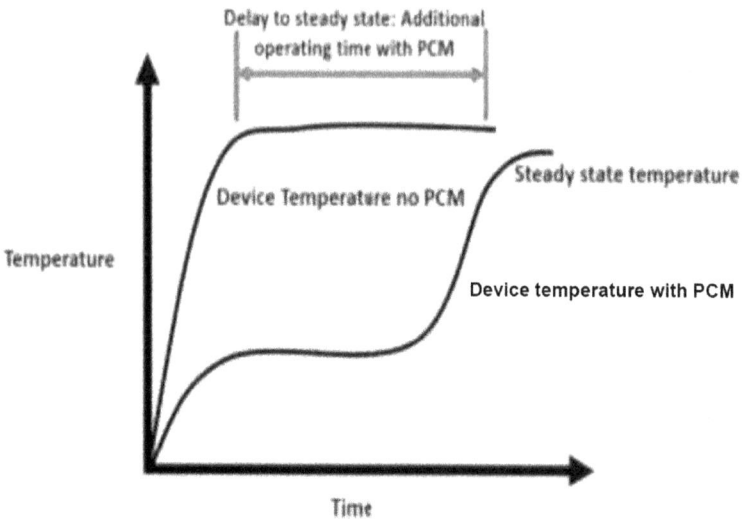

FIGURE 15.1   Melting process of PCMs[15].

15.3.3 FOOD INDUSTRY

As heat storage and transit systems, PCMs have been evaluated in food processing sections, chilling storage units, and packaging applications. In the drying of sweet potatoes, paraffin wax has been found to be energy-efficient. Heat pipe prototypes employing PCMs have been shown to significantly enhance food temperature distribution. Rubitherm RT6, filled with nanoscale calcium silicate (NCS), kept the container temperature at 10°C for 5 h while the temperature outside increased. Food-related smart packaging solutions could incorporate micro/nanostructures that contain dodecane and zein. The electrospinning technology was utilized to increase the heat storage capacity of polylactide (PLA), polycaprolactone (PCL), and dodecane in another investigation. As much as 34% of PCMs can be modified by encapsulating RT5, according to the results. Thermal energy savings and improved thermal processes (e.g. food supercooling applications and chilling bins) are two additional benefits of PCM use.

15.3.4 BUILDINGS

Energy storage and energy efficiency in buildings are of major importance since buildings utilize 30–40% of the entire amount of energy produced (such as space-conditioning). PCMs can be utilized in a variety of construction applications, including ceilings, floors, walls, and HVAC (heating, ventilation, and air conditioning) systems. Also, PCM-containing building materials can store energy. Nanocomposite materials are made by mixing PCM-Al powder with epoxy resin in a poly ethylene glycol (PEG) resin to increase the thermal performance of buildings. And it was found that polyethylene terephthalate/polyethylene eutectics electrospunfibers might be used for energy storage in building constructions. Nearly 79% less energy is needed to maintain a room's temperature within the thermal comfort zone when nanoPCMs are included into the drywall. Thermal characteristics have also been improved by the use of hybrid nanocomposite PCMs[15].

15.3.5 SOLAR ENERGY STORAGE

Solar hot water systems (Figure 15.2) with flat solar collectors installed on the roof feature a heat transfer liquid running through the receiver tubes. Solar power systems are cost-effective and reduce fossil fuel consumption. Using a geothermal heating and cooling system, for example, PCMs could allow solar power systems to be used at night even when there is little or no solar energy available.

TES system medium mass and necessary overall tank volume are reduced by about 30% and 65% using PCM-based systems, according to the results. Polyethylene terephthalate is the polymer matrix while LA fibers are the PCM. In the solar water heater, paraffin wax and a nanocomposite of paraffin wax with 1% nano Cu particles are used as the energy storage medium. The thermal conductivity of nanocomposites was enhanced by up to 24% over pure paraffin wax. Adding CuO NPs to coconut oil PCMs improves the melting process.

Solar collector yields could be improved with the use of many types of low and high molecular-weight nanomaterials, including nanofluids, PCMs, and nanocomposites.

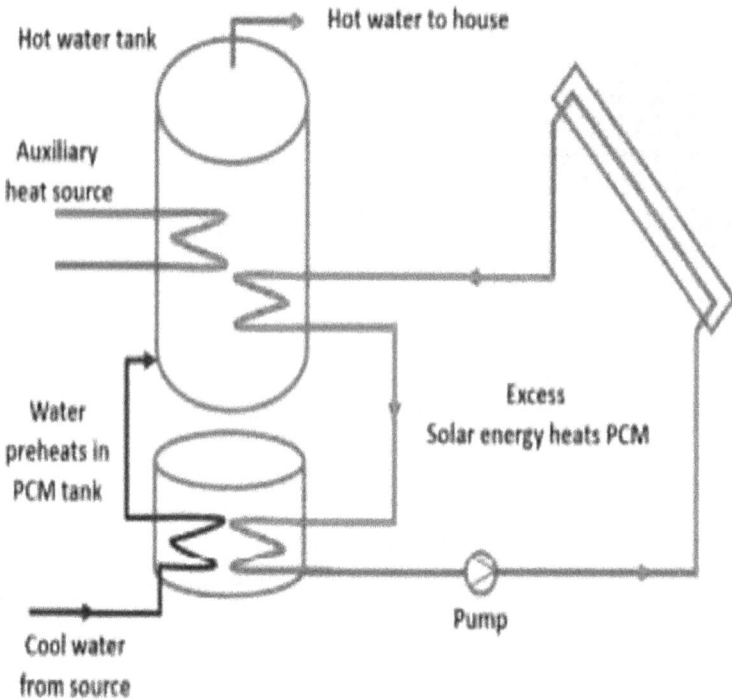

FIGURE 15.2   Solar hot water system.

Capsules of 600–900 nm in diameter made of Cu and Cu_2O and paraffin as the core by in-situ polymerization have showed excellent LHS capacity and thermal conductivity of about 128.55 J gl and 0.92 W/m K respectively[15].

15.3.6   HEAT EXCHANGERS

Heating systems can benefit from the increased efficiency and storage capacity provided by PCMs. As a result of this, they also reduce the size of pipelines, heat exchangers, and transportation energy consumption. In Figure 15.3, a shell and tube heat exchanger system using PCMs is depicted. The tube side of the PCM can, however, be utilized in some design concepts. Heat exchangers with PCM-graphite in a hot water tank are capable of storing roughly 3% more energy than a water-only tank. The low thermal resistance between the PCMs in the shell and the heat transfer fluid in the tubes is fundamental to the development of PCMs for TES.

PCM charging times may be slashed by incorporating internal tube fins into PCMs. Methods for designing heat exchangers that improve thermal response are available. Consequently, PCMs with various melting points, high conductivity particles incorporated, and fins have been produced. PCMs based on silver NPs could significantly improve the thermal conductivity of an air conditioning system. A multitube heat exchanger employing RT50 as the PCM was also shown to improve the solidification rate by increasing the concentration of Cu nanoparticles[15].

Heat transfer fluid passes
through each tube

PCM fills the shell

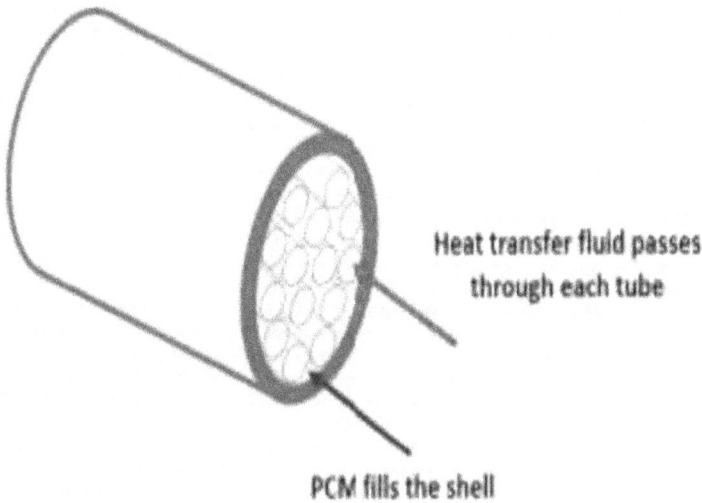

FIGURE 15.3   PCM-based heat exchanger system.

15.3.7 TEXTILES

PCMs are also employed in the development of various textiles, clothing, and footwear for the purpose of thermal management. In heated situations, the PCMs, for example, can absorb heat, making the operator more at ease. PCMs can be directly incorporated into fabrics' fibers, addressing the issue of coating durability. Nanosilver-coated nonadecane is used to increase the thermo-regulating properties of cotton garments and is coated with cotton. Using PCMs on the outer layer of clothes could improve its heat-regulating properties, according to the study's findings. Coaxial electrospinning has also yielded ultra-fine PCFs based on PEG and CA.

The results showed that the ultimate strain and strength of the compound fibers were lower than those of CA fibers. It has also created nanofibers based on PEG (PEG600 and PEG1000) as PCMs and PVDF as a supporting polymer using single and coaxial electrospinning techniques. A single electrospinning technique was employed to avoid altering nanofiber shape and allowing PEG to leak. Nanocapsules with a diameter of 141 nm were subsequently coated on the fabric using the pad-dry-cure technique, demonstrating that the treated cloth with nanocapsules had a high thermal stability[15].

15.3.8 PACKED BED DESIGNS

Large spherical capsules with PCMs are put into the tank and the heat transfer fluid flows through all parts (Figure 15.4). More heat exchange contact area and lower equipment costs are only a few advantages of packed bed systems. Thermal conductivity of the capsule materials was shown to have a negligible effect on the discharging performance of encapsulated thermal storage tanks.

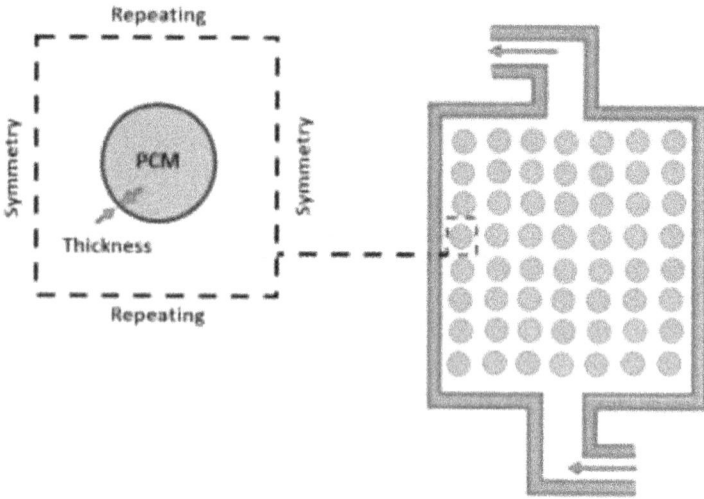

FIGURE 15.4 Application of PCM in packed bed design[15].

15.3.9 SPACE SYSTEMS

Due to the extreme temperature fluctuations in space, the PCMs can also be used for a high level of thermal control for space systems. The thermal capacitors in NASA missions are some forms of the finned internal zones filled with PCMs (such as paraffin) and an external housing (Figure 15.5) which are used for thermal control. When payload transferred from the transport vehicle to the International Space Station, it can be unpowered about 6 h that the PCMs release thermal energy to maintain the payload warmth.

The effect of magnetic fields on PCMs: In the physical concept, a magnetic field refers to the field that transmits the magnetic effect between objects. The existence

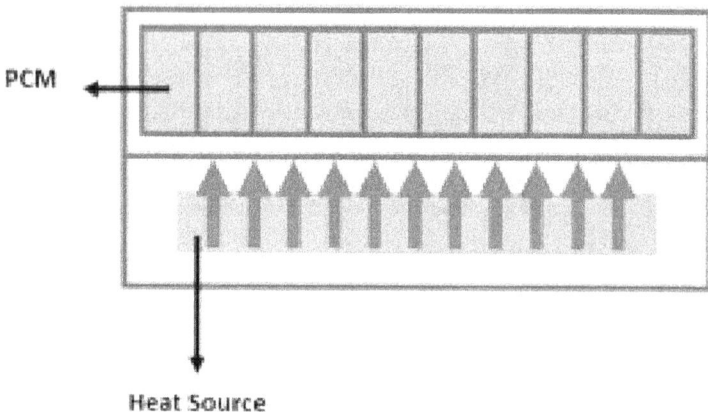

FIGURE 15.5 Application of PCM inthermal control systems[15].

of magnetic field can control the flow and heat transfer through Lorentz and Kelvin forces.

The effect of magnetic intensity, type, and location on melting and solidification: The magnetic field effect on the melting and crystallization of metals is examined. The shorter melting time and higher nucleation are still unclear. Therefore, the precise control of melting/solidification by magnetic field is still unknown and needs further studies. The simultaneous effects of two inhomogeneous magnetic fields on the melting phase change behavior of a PCM-filled cavity were researched. The results showed that the development of the melting front entirely depends on the intensity ratio of two inhomogeneous magnetic fields.

When the intensity is 4.6 Gs, it held obvious promotion at the phase change stage. An external magnetic field is used to stimulate the unsteady solidification process. The results indicated that with increase of the Hartmann number from 0 to 10, the solidification time was shortened by 23.5% on average. The influence of different magnetic field types and intensities on the freezing process is studied. The permanent magnetic field (PMF) and the alternating magnetic field (AMF) were compared. The shorter freezing time and lower energy needed made PMF apromising method for the higher freezing quality. The location of the magnetic field also has a different influence on phase transition.

The different location of coils leads to different influence on the melting process. When coils are on the right side, the magnetic force would result in suppression of the melting. However, when coils are on the bottom or left side, the melting process would both be enhanced. Figure 15.6 shows the comparison of the melting volume fraction when coils are on different sides[15].

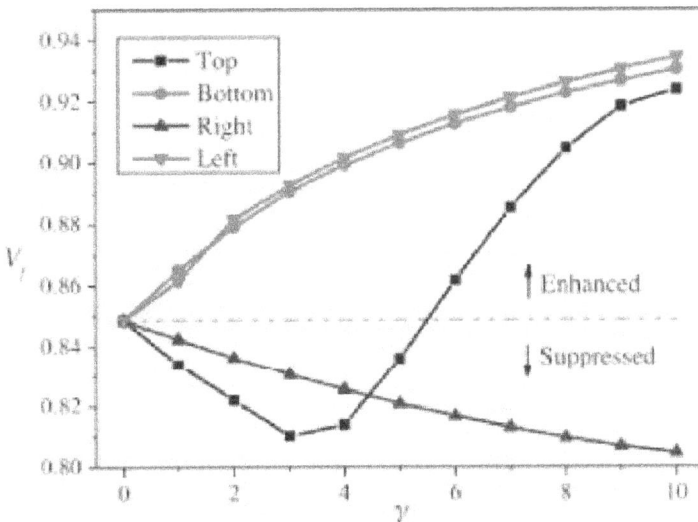

FIGURE 15.6 Comparison of the melting volume fraction when coils are on different sides[15].

The effect of the magnetic field on natural convection and boiling heat transfer: A numerical study is presented for natural convection flow of electrically conducting liquid gallium in a square cavity. The results show that the magnetic field with inclined angle has effects on the flow and heat transfer rates in the cavity. The analysis of melting behavior of PCMs in a cavity subject to a non-uniform magnetic field also confirmed that the magnitude and location of the magnetic field source do not show a significant effect on the natural convection heat transfer.

The existence of an intensive magnetic field accelerates the melting process and enhances the convection heat transfer. Applications of PCM in external fields are shown in Figures 15.7–15.12. The existing researches are all about the

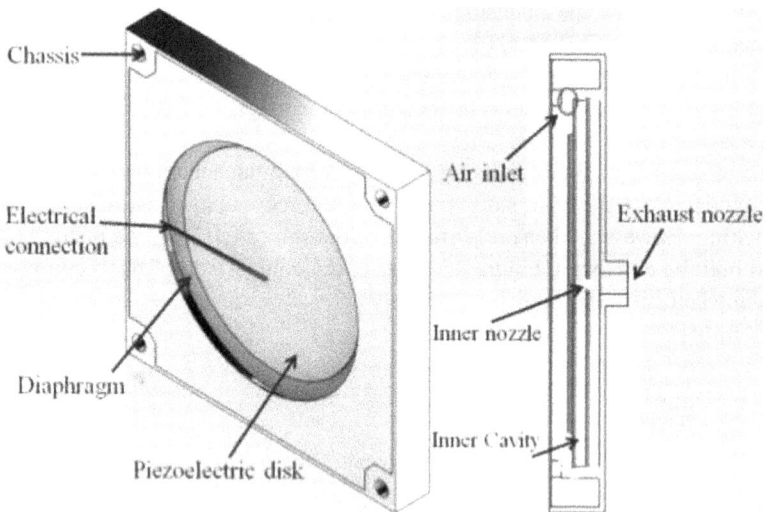

FIGURE 15.7 Ultrasonic micro-blower device for electronics cooling applications[16].

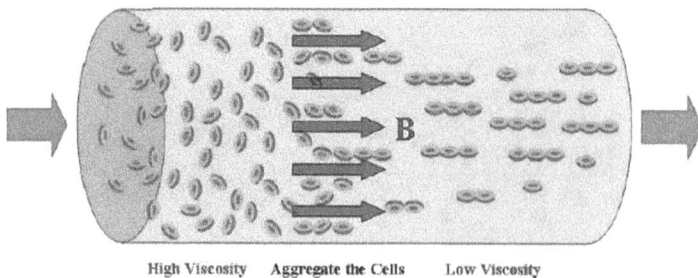

High Viscosity   Aggregate the Cells   Low Viscosity

FIGURE 15.8 Magnetic field for blood viscosity[17].

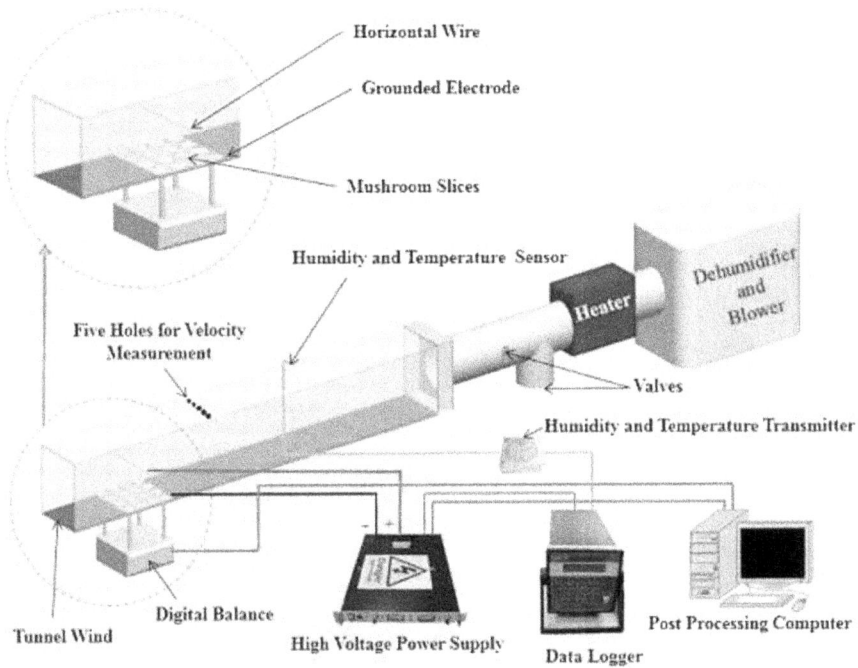

FIGURE 15.9 Combined convective electrodynamic drying system[18].

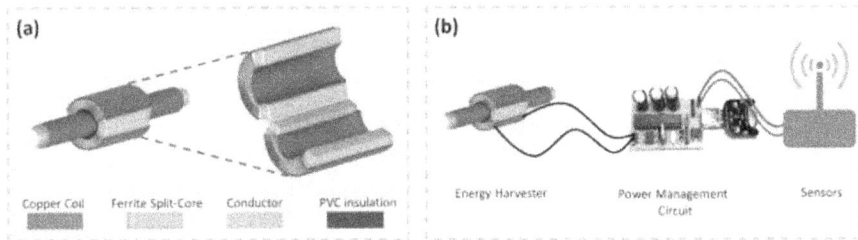

FIGURE 15.10 An indoor power line based magnetic field energy harvester[19].

influence of internal modification on phase change materials, but the influence of external factors on phase transition is still rare because of the wide application prospects mentioned earlier. This review entails a comprehensive insighton the effect of different external fields such as electric fields, magneticfields, ultrasonic waves, mechanical vibration on phase transition process, and heat transfer enhancement[16–20].

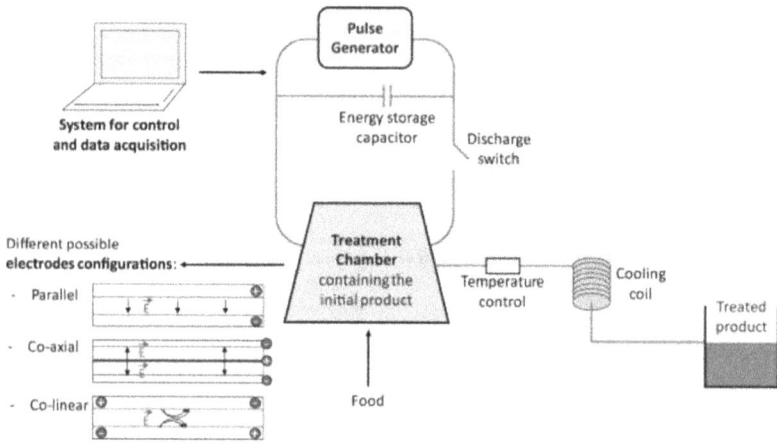

FIGURE 15.11 Pulsed electric fields in meat and fish processing industries[20].

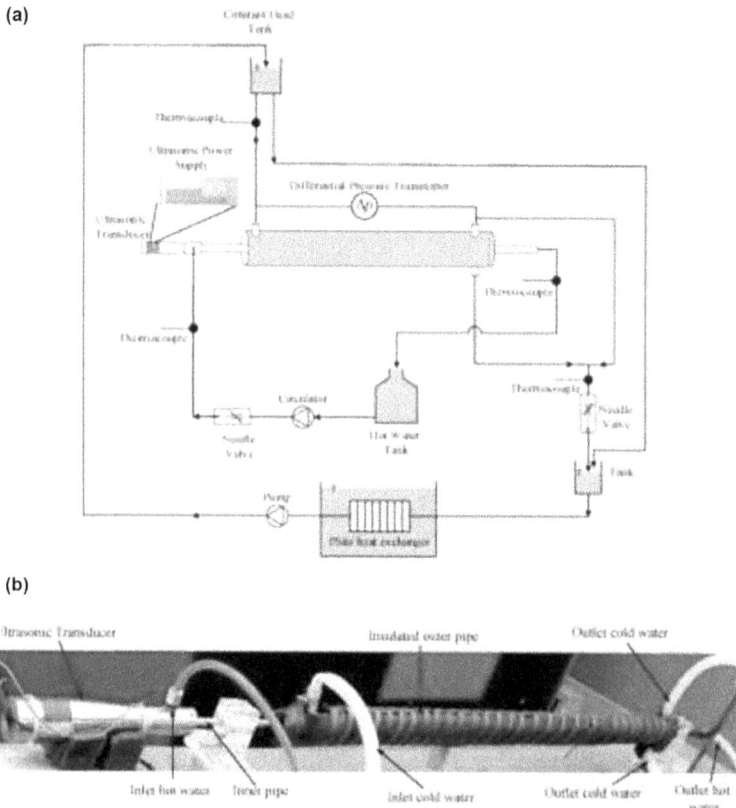

FIGURE 15.12 A double pipe heat exchanger using ultrasonic vibration[20].

15.4   SUMMARY OF EXTERNAL FIELD APPLICATIONS

Low noise and high heat transfer coefficient.

Reducing the blood viscosity by 20%–30%.

High drying rate and low energy consumption.

A sustainable and rechargeable free power supply for self-powered wireless applications.

Improving preservation, tenderization, aging, and increasing water holding properties.

Heat transfer enhances about 60% for both cold and heat fluids.

REFERENCES

[1] Wong, K. V. & De Leon, O., 'Applications of nanofluids: Current and future', *Advances in Mechanical Engineering*, vol. 2, p. 519659, 2010.

[2] Tyler, T., Shenderova, O., Cunningham, G., Walsh, J., Drobnik, J. & McGuire, G., 'Thermal transport properties of diamondbasednanofluids and nanocomposites', *Diamond and Related Materials*, vol. 15, nos. 11–12, pp. 2078–2081, 2006.

[3] Das, S. K., Choi, S. U. S. & Patel, H. E., 'Heat transfer in nanofluids: A review', *Heat Transfer Engineering*, vol. 27, no. 10, pp. 3–19, 2006.

[4] Liu, M.-S., Lin, M. C.-C., Huang, I.-T. & Wang, C.-C., 'Enhancement of thermal conductivity with carbon nanotube for nanofluids', *International Communications in Heat and Mass Transfer*, vol. 32, no. 9, pp. 1202–1210, 2005.

[5] Choi, S. U. S., Zhang, Z. G. & Keblinski, P., 'Nanofluids', In: Nalwa, H. S. (Ed.), *Encyclopedia of Nanoscience and Nanotechnology*, vol. 6, Los Angeles, CA: American Scientific, pp. 757–773, 2004.

[6] Murshed, S. M. S., Tan, S.-H. & Nguyen, N.-T., 'Temperature dependence of interfacial properties and viscosity of nanofluids for droplet-based microfluidics', *Journal of Physics D*, vol. 41, no. 8, Article ID 085502, 5 pages, 2008.

[7] Wong, K.-F. V. & Kurma, T., 'Transport properties of alumina nanofluids', *Nanotechnology*, vol. 19, no. 34, Article ID 345702, 8 pages, 2008.

[8] Wong, K.-F. V., Bon, B. L., Vu, S. & Samedi, S., 'Study of nanofluid natural convection phenomena in rectangular enclosures', In: *Proceedings of the ASME International Mechanical Engineering Congress and Exposition (IMECE '07)*, vol. 6, Seattle, Wash, USA, pp. 3–13, Nov. 2007.

[9] Ju-Nam, Y. & Lead, J. R., 'Manufactured nanoparticles: An overview of their chemistry, interactions and potential environmental implications', *Science of the Total Environment*, vol. 400, nos. 1–3, pp. 396–414, 2008.

[10] Routbort, J., et al., Argonne National Lab, Michellin North America, St. Gobain Corp., 2009. www1.eere.energy.gov/industry/nanomanufacturing/pdfs/nanofluidsindustrial cooling.pdf.

[11] Han, Z. H., Cao, F. Y. & Yang, B., 'Synthesis and thermal characterization of phase-changeable indium/polyalphaolefinnanofluids', *Applied Physics Letters*, vol. 92, no. 24, Article ID 243104, 3 pages, 2008.

[12] Donzelli, G., Cerbino, R. & Vailati, A. 'Bistable heat transfer in a nanofluid', *Physical Review Letters*, vol. 102, no. 10, Article ID 104503, 4 pages, 2009.

[13] Kim, S. J., Bang, I. C., Buongiorno, J. & Hu, L. W. 'Study of pool boiling and critical heat flux enhancement in nanofluids', *Bulletin of the Polish Academy of Sciences: Technical Sciences*, vol. 55, no. 2, 2007.

[14] Sujith, S. V., Kim, H. & Lee, J., 'A review on thermophysical property assessment of metal oxide-based nanofluids: Industrial perspectives', *Metals*, vol., no. 1, p. 165, 2022.

[15] Alehosseini, E. & Jafari, S. M., 'Nanoencapsulation of phase change materials (PCMs) and their applications in various fields for energy storage and management', *Advances in Colloid and Interface Science*, p. 102226, 2020.

[16] Ghaffari, O., Solovitz, S. A., Ikhlaq, M. & Arik, M., 'An investigation into flow and heat transfer of an ultrasonic micro-blower device for electronics cooling applications', *Applied Thermal Engineering*, vol. 106, pp. 881–889, 2016. www.physicscentral.com/buzz/blog/index.cfm?postid=4943264455129080433 (Access on 12th March, 2022).

[17] Johnson, M. J. & Go, D. B., 'Recent advances in electrohydrodynamic pumps operated by ionic winds: A review', *Plasma Sources Science and Technology*, vol. 26, no. 10, p. 103002, 2017.

[18] Maharjan, P., Salauddin, M., Cho, H. & Park, J. Y., 'An indoor power line based magnetic field energy harvester for self-powered wireless sensors in smart home applications', *Applied Energy*, vol. 232, pp. 398–408, 2018.

[19] Gómez, B., Munekata, P. E., Gavahian, M., Barba, F. J., Martí-Quijal, F. J., Bolumar, T., Campagnol, P. C. B., Tomasevic, I. & Lorenzo, J. M., 'Application of pulsed electric fields in meat and fish processing industries: An overview', *Food Research International*, vol. 123, pp. 95–105, 2019.

[20] Setareh, M., Saffar-Avval, M. & Abdullah, A., 'Experimental and numerical study on heat transfer enhancement using ultrasonic vibration in a double-pipe heat exchanger', *Applied Thermal Engineering*, vol. 159, p. 113867, 2019.

Index

For Product Safety Concerns and Information please contact our EU
representative GPSR@taylorandfrancis.com
Taylor & Francis Verlag GmbH, Kaufingerstraße 24, 80331 München, Germany